BASIC
DRAFTING

BASIC
DRAFTING
BY MARTIN CLIFFORD

TAB BOOKS Inc.
BLUE RIDGE SUMMIT, PA. 17214

FIRST EDITION

SECOND PRINTING—DECEMBER 1980

Copyright © 1980 by TAB BOOKS, Inc.

Printed in the United States of America

Library of Congress Cataloging in Publication Data

Clifford, Martin, 1910-
  Basic drafting.

  Includes index.
  1. Mechanical drawing. I. Title.
T353.C595   604'.2   79-25257
ISBN 0-8306-9945-7
ISBN 0-8306-1202-5 pbk.

Introduction

Drafting is a pictorial method of conveying instructions. In many respects drafting is superior to verbal instructions. While these instructions may be misunderstood or misinterpreted, a drafting plate is precise, not exact. The difference lies in the fact that the person using the drafting plate is often allowed some leeway.

Drafting is an international language. A drafting plate produced in one country can easily be understood in another. That old proverb, generally attributed to the Chinese, that "a picture is worth a thousand words," has basis in fact. It is true in the case of drafting.

One of the more useful arts is drafting. It can be considered on a par with arithmetic, reading or writing. Even if you have no intention of ever becoming a professional draftsman, learning drafting will make you more observant and will stimulate your thinking processes. Drafting is useful to hobbyists, for it is a sound basis for all types of construction projects. It is a useful technique for inventors, technicians and engineers of all kinds because it enables them to put their ideas on paper.

Drafting is used in just about every industry you can name. You will find drafting plates in plumbing, heating, building, architecture, electrical work, road building and in machine making.

This is an elementary book on drafting. It assumes you know little or nothing about the subject but want to learn. Learning drafting involves more than just reading about it. You must practice drafting as much as you can. Practice takes time and effort.

This book will not make you into a top-notch professional draftsman. Students who graduate from medical school do not start out as brain surgeons, either. But it is a start, and it will be a base on which you can build toward a professional career, should you wish to do so.

Even if you have no wish to do any drafting and have no intention of ever making a drawing, the ability to read and understand a drawing is comparable with the ability to read and understand a written sentence. Drawings are used for hobbies, business and industry. Being able to read a drawing makes each of these functions more enjoyable and profitable.

Martin Clifford

TO ADRIENNE

with respect, admiration and love
(but not in that order)

Contents

Chapter 1
Drafting Equipment

As in any other profession, drafting has its own tools of the trade. Some draftsmen do spend hundreds of dollars for specialized equipment, generally paying for it out of earnings, but there is no need to do this right at the start. You can buy enough drafting materials to get you launched on what could be a lifetime career or as an aid to a hobby without breaking your budget.

There is an almost endless variety of tools you can use for drafting. Some of these are highly specialized, being designed for such particular jobs as marine and aviation drafting; others can be used for almost any form of drafting. The drafting board, drafting table, protractors, triangles, compasses, French curves, dividers, ruling pens and bow instruments are among those you will probably need most often.

THE WORK SURFACE

All drawings are produced on a *work surface*. This can be as simple as an inexpensive drafting board, or it can be the more expensive drafting table or the still more costly drafting machine (Fig. 1-1).

You will find a wide assortment of drafting tables in all sizes and price ranges. The area of the work surface depends on the size of the drawings that are to be made. Some drawings are very large and must have a larger working area.

The Drafting Board

A *drafting board* is not only an excellent area for beginning drawings, but, because it is light weight and portable, it can also be used as a

Fig. 1-1. Basic drawing board. It has the advantages of convenience, portability and economy.

supplement for fixed-position tables. The drawing board (or drafting board) is generally made of white pine or some other soft wood. The important features of a drawing board are that its surface should be flat and smooth, not contain any broken areas and that it should be absolutely square. This means that the corners must form right angles with each other. It is particularly important that the left side has this 90° angle with the rest of the board throughout its entire length. The basic board is just that—a board; but some drawing boards have attachments available so that the board can be used flat, tilted or lifted at one end.

The Drafting Table

The *drafting table*, like the drawing board, is a working surface. There are many different kinds of drafting tables, but they all have two basic factors in common. One is that the surface is absolutely flat and unbroken; the other is that the board must be square. Larger tables have no legs but are hydraulically adjustable; others have telescoping legs.

In the table illustrated in Fig. 1-2 the drawing surface of the table can be lifted or lowered by a raising rod at the rear of the top. The front end of the table is hinged. This table has a drawer for storing drafting equipment and a file drawer for holding various sizes of drawing paper. Completed drawings are generally stored in a separate metal file cabinet designed just for that purpose.

The drafting table shown in Fig. 1-3 isn't as elaborate or as costly as the one illustrated in Fig. 1-2, but it can be considered the workhorse of the drafting industry. It is one of the most commonly used. The top of the table can be easily adjusted and made to move from

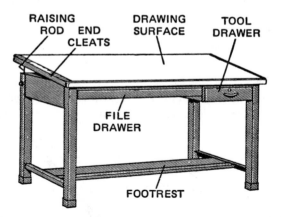

RAISING   DRAWING   TOOL
ROD   END   SURFACE   DRAWER
CLEATS

FILE
DRAWER

FOOTREST

Fig. 1-2. This type of drafting table has a large working area. It is often covered with a sheet of drawing paper to provide a smoother working surface, particularly if the table is old and its top area is nicked and cut.

Fig. 1-3. This is one of the most commonly used drafting tables. The working surface can be horizontal or tilted. The triple leg arrangement results in a table that is very stable.

the flat or horizontal position to any angle of tilt. This is done by a large screw with a substantial circular handle mounted directly beneath the top of the table. The bottom edge of the work surface has a lip to keep tools from sliding off the table. Unlike the previous table, this one does not have tool or file drawers. Clamp-on lamps, spring-loaded, can be mounted at the rear of the table to provide more light than that supplied by room overhead lights.

The Light Table

Still another type of drafting table is known as a *light table* and is illustrated in Fig. 1-4. This table has a glass working surface with fluorescent lights placed beneath it. It is used when it is necessary to trace other drawings or to reveal drawing details that could not

Fig. 1-4. A drafting table equipped with an illuminated glass working surface.

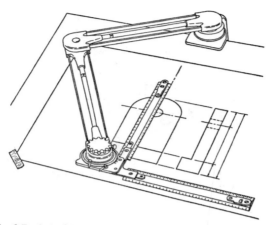

Fig. 1-5. A drafting machine is a precise tool for professionals.

otherwise be seen. Although the unit shown in Fig. 1-4 is a full table size, it is also available as a drawing board unit that is easily moved about and which can be placed on an ordinary table. Some light tables are made completely of metal and are not designed for drawing purposes, but rather to examine drawing details.

Drafting Machine

A drafting machine is a combination T-square, drafting triangle and protractor (Fig. 1-5). The T-square section is actually a pair of scales, so that you can scale as you draw, without the need for using external scales. The drafting machine, though, is expensive compared to a basic drafting kit, and is used by professional draftsmen. It isn't designed as a drafting tool for those who are just beginning.

Cleanliness

It is good practice, when work is completed for the day, to cover drawing boards and tables to insure drawing cleanliness and to protect drawings against possible damage. The simplest cover is a sheet of drawing paper, somewhat larger than the drawing on the board. It can be held in place with several pushpins. The subject of cleanliness is an important one, and will be covered in more detail in the next chapter when you start work on your first drawing.

Light

More light is needed for drafting than for most other types of indoor jobs. You will find that drafting rooms often have much more illumination than rooms used for other kinds of work. Even so, many draftsmen prefer attaching a lamp to the rear of the drawing board. Since this is a swivel type, it can be positioned in any manner so different areas of a drawing can be spotlighted.

Most of the fixtures use fluorescent lights. These have the advantage that they spread the light over a considerable area and that they operate cooler than ordinary light bulbs. However, they can flicker, but even the smallest amount of flicker will produce eye strain and possibly a headache. To get the best light, use a fixture that will hold two fluorescents. This tends to reduce the average amount of flicker, producing a more uniform light. Also, you can intermix "warm" and "cool" lights if you wish to produce the kind of light you find most pleasing.

DRAFTING TOOLS

You can start drafting with nothing more than a very simple kit of drafting tools, a T-square, a drawing board, a few triangles and some drawing pencils. You can always add to this basic setup, buying additional drawing tools as you need them. Figure 1-6 illustrates some of the drafting equipment that is available, but by no means all. Having a large amount of equipment is no assurance that the owner is a professional draftsman or has professional skill. Quality and sufficient equipment does mean it will be easier for a skilled draftsman to do his work.

T-Squares

A *T-square* is a drafting tool used in conjunction with a drawing board or table. The purpose of a T-square is to draw horizontal lines, to act as a support for triangles for drawing vertical and slant lines, and as a guide for the correct mounting of work sheets. The T-square gets its name from the way it is shaped. It consists of a long, relatively narrow strip called the blade, mounted at right angles to a much shorter strip referred to as the head. Figure 1-7 shows a drawing board with a T-square in position.

You will find T-squares of all sorts. If you are starting with a small portable board, then there wouldn't be much justification for buying a T-square so long that its end will hang well over the edge of the board. Some T-squares come equipped with a protractor head, a head that is adjustable to various angular positions. As a general rule, draftsmen seem to prefer T-squares having fixed heads.

If you wish, you can get a T-square with one or more scales engraved on the blade. And you can also get them in just about any length that will fit a drawing board.

The head of the T-square is mounted under the blade so that it fits against the edge of the drawing board while the blade rests on the

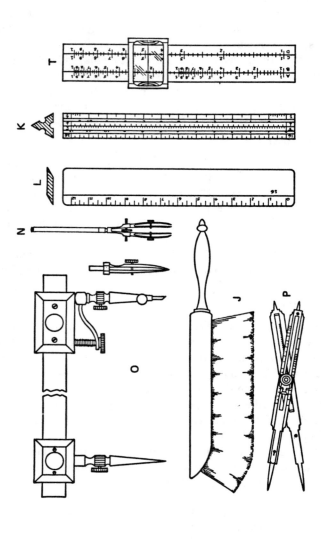

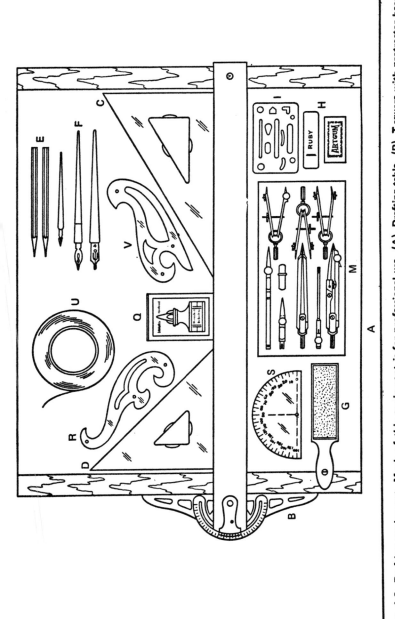

Fig. 1-6. Drafting equipment. Much of this equipment is for professional use. (A). Drafting table. (B). T-square with protractor head. (C). A 30°-60° triangle. (D). A 45° triangle. (E). Drawing pencils. (F). Lettering pens. (G). Pencil point shaper. (H). Erasing shield. (I). Erasers. (J). Dust brush. (K). Architects' scale. (L). A 12-inch scale. (M). Set of drafting instruments. (N). Road pen. (O). Beam compass. (P). Proportional dividers. (Q). Drawing ink. (R). French curve. (S). Protractor. (T). Slide rule. (U). Spool of paper fastener tape. (V). French curve.

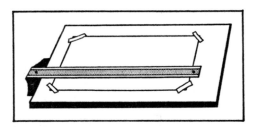

Fig. 1-7. Paper is fastened to board and the T-square is in position.

surface. T-squares vary in length from 15 inches to as much as 72 inches. The 36-inch length is the most common. T-squares are made of wood, a combination of wood and plastic or of metal. The most important feature of a T-square is that the head and the blade must absolutely be at perfect right angles to each other. Any deviation will result in drawings in which the lines will be "out of square."

Some T-squares are made so that the head and blade are held together by screws, while in others the two components of the T-square are fastened together permanently. You can check a T-square in many ways. One technique is to draw a right angle, using the T-square, and then to measure the angle with the help of a protractor. An easier way is to put a triangle right up against the T-square to see if the head and the blade form the proper 90° angle.

One of the basic rules of the T-square is that it must be kept clean. Keep a soft cloth handy for wiping the blade just to make sure it hasn't picked up any pencil markings or crumbs of erased material. Some drawings are done only in pencil; others are in ink. Both can contribute to a dirty T-square.

Handle the T-square with respect. If dropped, the head and the blade may no longer be at right angles to each other. Don't use the T-square as a guide for a razor blade when cutting paper. Use a cutting board for that. When you are finished with your work, don't leave the T-square on the work surface. Instead, use the hole at the end of the T-square as a support for holding the T-square suspended vertically from a nail on a nearby wall.

Before beginning any new work, test the top of the T-square for warps or nicks. Draw a sharp line along the top edge of the blade. Turn the T-square over and redraw the same line with the same edge. If the blade is warped, the two lines will not coincide. An easier way is to run your thumb along the blade edge. You should be able to feel any imperfections.

Lines drawn with the T-square must be parallel to the bottom and top edges of the work sheet. But this makes certain assumptions. You are assuming that the work sheet is square. This is usually the case, except for sheets that are damaged or warped. You are also assuming that the T-square has its head and blade perfectly aligned. And it is also possible, especially when working in a hurry, to mount the work sheet at a slight angle.

Finally, do not use the T-square for anything else. It isn't a ruler. It isn't supposed to substitute for a triangle. And it's not supposed to be used for pushing or shoving other objects.

The Parallel Straightedge

For large drawings, many draftsmen prefer a tool known as a *parallel straightedge*. The straightedge is permanently attached to the drawing board by a system of cords and pulleys so arranged as to give exact parallel motion to the straightedge as it is moved over the board.

The parallel straightedge has two fundamental advantages over the T-square (Fig. 1-8). Supported at both ends, it maintains its parallel motion automatically. It can be moved up or down the board with pressure at any point along its length. These advantages are particularly significant when working on large drawings because a T-square tends to become unwieldy and inaccurate when you work near the blade end of a long T-square.

Steel straightedges are used for drawing long straight lines. Sometimes a long T-square blade is used as a straightedge, but the steel straightedge is superior because its heavier weight helps keep the straightedge in position. Putting tape at both ends of the straightedge helps reduce unwanted movement. Using the beveled edge for inking reduces the chance of ink running under the straightedge. The beveled edge is the slanted edge.

Fig. 1-8. Technique for using a straightedge.

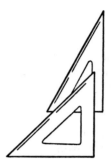

Fig. 1-9. Two of the three basic triangles used in drafting. The lower is a 45° triangle and the upper is a 60° triangle.

Triangles

A *triangle* is still another basic drafting tool. They are inexpensive, and are usually plastic, although you will find some made of metal. A triangle – any triangle – encloses a total of 180°. For any triangle, one of the angles is always 90°. The two remaining angles can be 45° and 45° or 30° and 60°. When the two angles are identical (45°), the unit is referred to as a 45° triangle. With the triangle held as in Fig. 1-9, the angle in the lower left corner identifies the triangle. If this is 30°, it is a 30° triangle, if 60°, a 60° triangle. There are various kinds of triangles, but the 30°, 45° and 60°, are the most common. Triangles are used for drawing vertical lines, slant lines and angles.

While triangles are fairly inexpensive, you should protect them for the simple reason that a defective triangle can ruin hours of work. Triangles can be easily supported by a nail since they have so much open space. It isn't advisable to store triangles in a drafting table drawer since such drawers often end up by being a storehouse of miscellaneous junk.

Triangles are plastic and can be deformed by heat, so it's a good idea not to put them where they can be reached by excessive warmth. Obviously, the top of a steam radiator is no place for them. If a triangle warps, put it on a flat surface and then pile a vertical stack of books on top of it. Leave it that way for several days. If it is still warped at the end of that time, replace it.

The Protractor

A *protractor* is a very simple drawing tool made in either circular or semicircular form as shown in Fig. 1-10. It is used for measuring angles. The outer edge is numbered in degrees, ranging from 0° to 180° for the semicircular type and up to 360° for the circular. This numbering arrangement is repeated on the protractor edge in reverse form, so that degrees can be measured from left to right or from right to left.

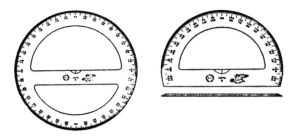

Fig. 1-10. Circular and semicircular protractors.

The Compass

No matter how simple or elaborate your drafting set may be, you will find a *compass* included in it. A compass is a drafting tool used for drawing complete circles or parts of circles known as arcs (Fig. 1-11). An arc is a portion of the circumference of a circle. Like other drafting instruments, the compass is available in a great variety of sizes and shapes and can be manufactured from a number of different metals. The compass shown in Fig. 1-12 can be opened or closed by turning a small wheel located between the legs of the compass.

A compass can be a pencil type or it may be designed for drawing in ink. On some compasses there is an interchangeability feature, so the same compass can be used for either pencil or ink drawings. Each compass also has a needle point, generally adjustable by a small screw. To set the compass properly, turn the center screw of the compass until both legs of the compass are next to each other. Then adjust the screw of the needle point until it projects about 1/32 inch beyond the pencil point or the extreme tip of the ink reservoir.

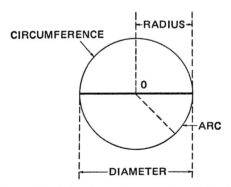

Fig. 1-11. Parts of a circle. An arc is any section of the circumference. A diameter is the maximum straight line distance inside the circle through center point O. A radius is equal to one-half the length of the diameter.

**Fig. 1-12. One type of compass. You can make adjustments
by turning the center wheel.**

It is best to start with a pencil compass. Your compass may come
equipped with an extension bar to be inserted between the main body
of the compass and the pencil or pen leg. Its purpose is to extend the
range of the compass to let you draw very large circles, but first practice using the compass without it. The average 6-inch compass can
draw circles up to about 12 inches in diameter, large enough for
most work drawings.

Dividers

A *divider* looks something like a compass, but unlike that instrument comes equipped with a pair of needle points. You can use dividers for making measurements, for transferring measurements, or
for dividing a line into equal numbers of parts.

Proportional dividers are used for the transfer of measurements
from one scale to another and are useful when drawings are to be
made to a larger or to a smaller scale. Figure 1-13 shows a typical
pair of proportional dividers.

The proportional dividers have legs which cross each other at a
pivot point which is adjustable. The lengths of the legs on either side
of the pivot can be made longer or smaller. If, for example, the distance between one pair of divider points is 1 inch and that between
the other pair is 1/2 inch, then the ratio is 2 to 1. This means you
can make a drawing which is either 1/2 the dimensions of some object
or other drawing, or one which has two times the original dimensions.
The example given here is that of a ratio of 2 to 1, but other ratios, such
as 3 to 1, 4 to 1, etc., are possible, including fractional proportions.

The section at which the two legs of a compass or a divider meet is
called a *joint*. Inexpensive drafting instruments usually have a tongue

Fig. 1-13. Proportional dividers.

Fig. 1-14. The tongue joint. You may find this type of joint on inexpensive drafting tools.

joint of the type illustrated in Fig. 1-14. The trouble with a tongue joint is that there is no way of tightening it after it has become worn. When this happens, the two legs become loose, making working with such tools more difficult. When you buy a divider or a compass having a tongue joint, make sure the legs are reasonably tight and that there is a minimum of play.

Better grade drafting instruments use pivot joints. Such tools cost more because they are more expensive to manufacture. They are available in a number of types, as shown in Fig. 1-15. These generally have tiny adjusting screws or pivot screws so the legs can be tightened should they become loose.

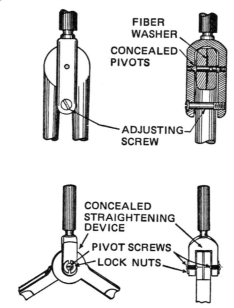

Fig. 1-15. Two types of joints used in better-grade instruments.

When you buy dividers, run your fingers along the full lengths of the needle points to make sure there are no burrs and that the needle point hasn't been damaged. Sometimes the very edge of the needle point will be bent in or may be blunted. For dividers, both legs must have the same length. You can check this by closing the instrument and resting the needle points on a hard surface. When doing this, the drawing tool will be absolutely vertical and both needle points should rest directly on the surface.

French Curves

The compass can be used to draw a circle or an arc, but you may also need to connect a series of points on a drawing by a smooth curve. For drawing such lines, you will find it helpful to use one of a series of guides called *French curves*. French curves, as illustrated in Fig. 1-16, are available in a variety of sizes and shapes and, like triangles, are made of opaque or see-through plastic.

Templates

A *template* is a device, generally made of plastic, designed to let you draw accurate symbols, such as architectural, plumbing, electrical and electronic symbols (Fig. 1-17A). Using a template has a number of advantages. The first, of course, is the great saving in time. Still another is that the symbols are always the same shape and size.

There are also templates available to let you draw a large number of different geometric shapes, such as triangles, arcs and ellipses. Templates come in different sizes and the larger ones, of course, have a correspondingly larger number of symbols. Some triangles, as in Fig. 1-17B, are combined triangles and templates. This template is for drawing small circles and hexagons.

Fig. 1-16. Various types of French curves.

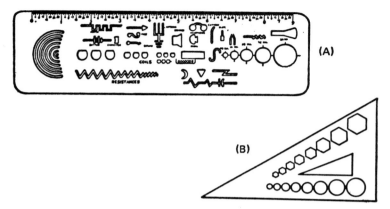

**Fig. 1-17. (A). Template for electrical symbols.
(B). Combined triangle and template.**

Bow Instruments

Drawings of circles with diameters of less than 1 inch are difficult to do with the ordinary compass. Instead, bow pens and bow pencils, as illustrated in Fig. 1-18, can be used. Bow dividers can also be used for transferring small measurements.

Bow instruments have a side-mounted screw which works against spring pressure for opening and maintaining the position of the instrument legs. The screw threads on this drawing instrument are delicate and should never be forced. When you increase or decrease the spread of the legs, hold them together lightly to decrease the tension on the setscrew and to decrease the wear on the threads. Keep the threads free of rust and dirt.

The drop bow pen isn't a standard instrument, for it is used to draw circles having diameters of less than 1/4 inch. It is usually included only with the more expensive drawing sets.

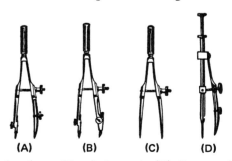

**Fig. 1-18. Various types of bow instruments. (A). Bow pen. (B). Bow pencil.
(C). Bow dividers. (D). Drop bow pen.**

TAKING CARE OF YOUR INSTRUMENTS

You can buy drafting tools individually or as a set in a case. There is no reason to buy an expensive set for a number of reasons. You don't need it for the practice work required to learn how to do drafting. It is also better to learn how to handle the less expensive variety first because drafting tools can be damaged if not used properly. Then there is also the matter of cleaning and storing drafting tools. Finally, after using an inexpensive set, you will really appreciate the quality and workmanship that goes into some drafting equipment.

To protect instruments when not in use, clean them frequently with a soft cloth. Scrap cloth is satisfactory, if it is thread and lint free — and reasonably clean.

From time to time, put a few drops of a very light oil on the cloth and rub your instruments. Do not oil any of the joints. This sort of rub will keep rust spots from forming and will act as a protective film against any corrosive substances that may be in the air. When all work requiring drafting instruments is completed, put the instruments in their case and not in a desk drawer. It's a good habit to wipe them before doing so since perspiration can affect the instruments.

Drafting tools aren't toys, and they shouldn't be used as such either by children or by adults who have a lot of curiosity and little consideration. It's not a bad idea to keep equipment under lock and key, even if it is a budget-type drafting set. You would be surprised at how much good work can be obtained with such tools with some effort and practice.

SCALES

Drawing an object to scale means that the lines of the drawing will have the same dimensions as the object. This doesn't present a problem when the object is small. When the item being drawn has dimensions in feet or yards, though, then it is obviously impractical to have a same size drawing. Same size drawings are made with a drafting technique known as *lofting*, but this is exceptional.

If a drawing is made half scale, every line is half the size of the dimensions of the actual object. On a drawing that is half scale, a line that is 1 inch long represents an actual dimension of 2 inches on the object.

Types of Scales

All measurements on drawings are made with *scales*. You can identify scales in two ways — by their shape and by their use.

Figure 1-19 shows a few of the more commonly used scales. The triangular unit shown at the top of Fig. 1-19 has six scales, two on each

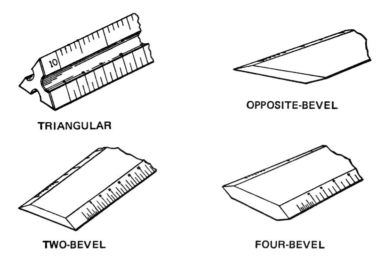

TRIANGULAR

OPPOSITE-BEVEL

TWO-BEVEL

FOUR-BEVEL

Fig. 1-19. Scales can be identified by shape.

face, but it is also available with more. The two-bevel scale supplies two scales with the advantage that the scales, appearing on a slant, are easy to read. The opposite-bevel scale has one edge cut away so it is always easy to lift from the drawing board. Unlike the triangular and two-bevel scales, the opposite bevel type has only one scale. The four-bevel scale, shown as the bottom illustration in Fig. 1-19, has four scales on one rule, one on each side of a bevel. The four-bevel scale is usually about 6 inches long so you can easily carry it with you in a pocket. The other scales, the triangular, two-bevel and opposite bevel, are 12 inches long.

Architects' Scale. The architects' scale belongs to the triangular rule family (Fig. 1-20). An architects' scale has 11 scales: 3/32, 3/16, 1/8, 1/4, 3/8, 3/4, 1/2, 1, 1 1/2, 3 and a foot scale calibrated in sixteenths of an inch and inches. Each division on the architects' scale is equivalent to 1 foot on the object being drawn. If, for example, you use the

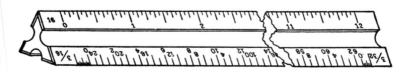

Fig. 1-20. Architects' scale has 11 scales.

1/8 scale and use it for measuring off 1/8 inch on your drawing, that 1/8 inch will be equivalent to 1 foot measured directly on the object.

The scales are divided into two areas. There are six scales that are read from the left end and five scales from the right. Architects' scales are used for making measurements on machine, architectural and structural drawings.

Engineers' Scale. The engineers' scale is still another member of the triangular rule family (Fig. 1-21). While the architects' scale uses the English system of measurement, the engineers' scale is divided decimally. Since the decimal system is based on a system of tens, the inch is divided into 10 parts. And so a tenth of an inch on a drawing could represent 1 foot on an object.

Metric Scale. The metric scale shown in Fig. 1-22 is a two-bevel type. This scale is calibrated in metric units, rather than the English scale (1/16 inch, 1/8 inch, 1/4 inch, 1/2 inch, etc.) or the decimal scale (inch separated into 10 equal divisions). The divisions on the metric scale are in centimeters and millimeters. If you measure an object using the metric system, you can use the metric scale for reduction purposes. Thus, 1 centimeter on the scale could be equal to 10 millimeters measured on the object.

The metric system is in wide use in many countries and is gradually being adopted in the United States. Table 1-1 shows the relationship between metric and English units.

Indicating The Scale

There are various ways of indicating scale on a drawing. In the fractional method, the scale is written as a fraction or ratio; 1/4 is

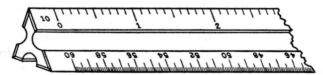

Fig. 1-21. Engineers' scale. Only a part of the scale is shown.

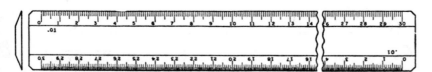

Fig. 1-22. The metric scale shown here is a two-bevel type.

Table 1-1. Metric/English conversion for linear measure.

1 millimeter	=	0.03937 inch
1 centimeter	=	0.3937 inch
1 meter	=	39.37 inches
1 meter	=	3.28083 feet
1 meter	=	1.0936 yards
1 kilometer	=	3,280.83 feet
1 kilometer	=	0.62137 mile
1 inch	=	25.4 millimeters
1 inch	=	2.54 centimeters
1 inch	=	0.0254 meter
1 foot	=	304.8 millimeters
1 foot	=	0.3048 meter
1 yard	=	0.9144 millimeter
1 mile	=	1.609 kilometers

a fraction and is also a ratio. Thinking of it as a fraction, it is 1 divided by 4. As a ratio it is the ratio of 1 to 4.

If you have a full-size scale drawing, the ratio is 1/1. This means the object and its drawing will have the same size. The object being drawn, though, can be smaller or larger than the drawing itself. The drawing unit is always the numerator; the object is always the denominator. In a ratio, such as 1/4, the digit 1 is the numerator. The digit 4 is the denominator. With an enlarged scale or when the drawing is larger than the object, the numerator, the first digit in the ratio, is always larger than the denominator, the second digit in the ratio. Thus, with a scale of 5/1, the drawing will be five times as large as the object. For enlarged scale drawings, the denominator is usually the digit 1, but not always. Thus, you could have a scale such as 5/2. This means that for every two units of the object, the drawing is 5 units. If a distance measured on the object is 2 inches, then the corresponding line on the drawing will be 5 inches.

For a reduced scale drawing, the numerator is always smaller than the denominator. On a 1/5 drawing, each section of the object will be five times as large as the corresponding section in the drawing. If a part of the object is 5 inches, then the line on the drawing representing that part will be 1 inch. As in the case of the enlarged drawing, a reduced drawing can have values other than 1 in the numerator. Thus, you could have 2/5 – or 3/10, etc.

Another method of indicating scale is known as the *equation method*. The symbol (") is used for inches and (') for feet; 6'5" means 6 feet 5 inches. A full scale drawing would be marked 12" = 1'. A reduced drawing might have its scale as 1/4" = 1'. This means that 1/4 inch on the drawing corresponds to 1 foot on the object. Or, for an enlarged scale, you could have 2' = 1". This means that 2 feet on the drawing is the same as 1 inch on the object. The first digit is the drawing measurement; the second digit is the object measurement.

PENCILS FOR DRAWING

You can group pencils into two categories: hard and soft. A soft pencil does not produce a very sharp line and has a tendency to smear. A hard pencil does give a sharp line, but can indent or cut the paper and may be difficult to erase. Further, the line it produces is very light and produces problems if a drawing is to be inked in over pencil lines. What we have here, then, are extremes, but draftsmen can obtain a large variety of pencils, depending on their own preferences.

There are 18 grades of pencil lead, ranging form 9H, the hardest, to 7B, the softest. The character of a pencil, its hardness or softness, is marked, stamped or printed somewhere on the wooden surface. Pencils can also be arranged in three categories in this way: 9H to 4H, hard; 3H, 2H, F, and HB, medium; and B to 7B, soft.

Choosing The Right Pencil

The character of a pencil depends not only on the personal preference of the draftsman, but also on the use to which the pencil is to be put. A soft pencil, for example, might be preferable for making rough sketches. A pencil having a medium lead is used for general purpose drawing, tracing and for lettering. The softer of the hard leads is used for line work in machine drawing. The harder lead is for graphs, charts or diagrams requiring a high degree of accuracy.

The standard leads described here are used primarily on paper and linen, although they may also be used on film. However, there now are leads made specifically for use on drafting film. The grade designations vary with the manufacturer; three of these are K1 through K5, 2S through 6S and E1 through E5.

To find the pencil that satisfies you best, you will need to experiment. If you want to keep your drawings clean, it would be best to stay away from the very soft pencils to avoid smearing. Try to choose a pencil that will make a smooth, continuous line, without breaks, and that will have uniform thickness along its length. Also, the pencil should not be so hard that you will be forced to press on it, possibly producing a groove in the work paper.

In drawing with a pencil, the width of the line you make will depend on the sharpness of the lead and not on the pressure you put on it. If the tip of the lead is very sharp, you can draw a thin, black line, or a thin, gray line, depending on the pressure of the pencil on the paper. Thus, pencil pressure changes line color, not line thickness   With a

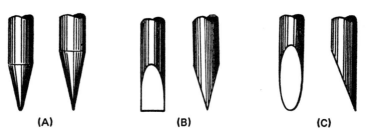

Fig. 1-23. Pencil point types. (A). Conical. (B). Chisel. (C). Elliptical.

slightly dull pencil tip, you can draw a medium black line or a medium gray line. The more blunt the tip, the wider the line you will draw.

Dressing The Pencil

Dressing a pencil means sharpening the pencil and preparing the point so it can be used on a drawing. Whether a wood pencil is used, or a mechanical clutch type which simply holds the lead, the lead must be dressed or pointed before it can be used (Fig. 1-23). The point can be conical, elliptical, or shaped like a chisel point. The conical is preferable, and you should use it for your first drawings.

To shape the pencil point, use a flat sandpaper pad as illustrated in Fig. 1-24. Rotate the pencil as you rub the tip on the pad, keeping the abrasive action of the pad even on all sides. The method for doing so is illustrated in Fig. 1-25. After you have sharpened the point to your satisfaction, test it by drawing a line on a sheet of scrap.

The trouble with the sandpaper pad is that it produces a lot of graphite dust. If you touch it with your fingers and then your drawing, you'll have a substantial mess to clean. A good technique is to mount the pad permanently on some flat, horizontal surface so you don't have to touch it with your fingers. Arrange the positioning of the pad so you can rotate the pencil point on it easily and conveniently. And when the sandpaper becomes filled with graphite particles, replace it, or you'll find pencils carrying quite a bit of black to your drawings.

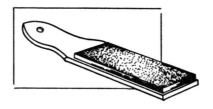

Fig. 1-24. Sandpaper pad mounted on a wooden handle.

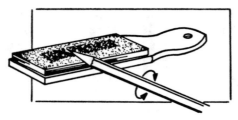

Fig. 1-25. Technique for producing a conical point using the sandpaper pad.

Fig. 1-26. Trim the pencil with a single-edge razor blade or a sharp penknife. Expose about 3/8-inch of lead.

To sharpen a drafting pencil, you can use a special pencil sharpener to cut the wood away from the lead without cutting the lead. It will leave about 3/8 inch of the lead exposed. Some draftsmen prefer a single edge razor blade or a sharp penknife (Fig. 1-26). Always sharpen that end of the pencil that is at the end opposite to the lettered designation of the pencil, so you will always know the type of pencil, no matter how often it is sharpened.

To avoid the need for pencil sharpening, some draftsmen prefer mechanical holders. These hold leads which, like pencils, are available from very soft to very hard.

THINGS TO DO

Based on what you have read in this first chapter, here are some practical work exercises:

- Using a pencil type compass, draw a large circle of any size.
- On this circle, draw two diameters. Draw a radius of the circle. Identify the circumference.
- Using a pair of dividers, measure a radius of this circle and then try to determine how it compares with the diameter.
- Using a single edge razor blade or a sharp penknife, sharpen a pencil so that the point is about 3/8-inch long.
- After sharpening a pencil try to produce these types of points: conical, chisel and elliptical.
- If you have any kind of template, practice with it by drawing all the symbols indicated on that template.
- Using your triangles, identify all the angles in those triangles.
- Examine all your drawing instruments, one by one, and identify each by name.
- Draw some curves at random using a French curve.

- Examine your protractor and become familiar with the way the numbers are indicated on it.

SUMMARY

- All drawings are produced on a work surface.
- Drawings can be done on a drafting board, a drafting table or a light table.
- A drafting machine is a combination T-square, drafting triangle and protractor.
- A T-square is used for drawing horizontal lines, for supporting triangles, and as an aid in the proper mounting of working sheets.
- A parallel straightedge is permanently attached to the drawing board by a system of cords and pulleys and supplies parallel motion automatically.
- A triangle is made of metal or plastic and is used for drawing vertical and slant lines.
- A protractor is made in circular or semicircular form and is used for measuring angles.
- A compass is a tool for drawing complete circles or arcs in either pencil or ink.
- Dividers resemble the compass but are equipped with a pair of needle points. They are used for making measurements, for transferring measurements or for dividing lines. Proportional dividers are used for the transfer of measurements from one scale to another.
- French curves, made of metal or plastic, are used for drawing curves.
- A template is a small rectangular section of metal or plastic with cutouts for drawing symbols.
- A bow instrument is a special type of compass for drawing small circles in pencil or ink. These circles are usually less than 1 inch in diameter. A drop bow pen is for drawing circles having a diameter of less than 1/4 inch.
- Various kinds of scales are used in drafting, including the two-bevel, opposite bevel, four bevel, triangular architects', triangular engineers' and metric.
- Pencils are designated as hard, medium and soft. Hard pencils are 9H to 4H; medium are 3H, 2H, F and HB, and soft are B to 7B.
- Dressing a pencil means sharpening the pencil and preparing the point for drawing use. Sandpaper pads are used for shaping pencil points.

Chapter 2
How to Use Drafting Tools

When first beginning drafting, it is better to work with a T-square and basic drawing board. This will enable you to get the "feel" of such tools and will also let you "square" the drawing paper on the board with less difficulty. For beginning work, use 8 1/2 x 11 worksheets.

PREPARATION FOR DRAWING

Before you begin your drawing work, put all the materials and equipment you may need; such as pencils, sandpaper pad, erasers, triangles and curves, within easy reach. It's good practice to sharpen a number of pencils in advance so you won't need to interrupt your work later on. A drawing can be completed faster (and neater) if you can work on it continuously.

You can do better work if your drawing board is covered with a sheet of firm paper. Firm means any paper that has a hard, smooth surface, preferably new, and not containing any holes, surface irregularities, or pencil markings. The reason for using a backing sheet is due to the fact that the wood of your drawing board or table may have pinholes or some slight unevenness because of the wood grain. The backing sheet will provide a smoother working surface.

When you put the covering sheet on your drawing board, be sure it does not stick over the edge on which your T-square slides. This is generally the left-hand side of your drawing board. The covering sheet (or backing sheet) should not interfere in any way with the free movement of your T-square.

CLEANLINESS

Your drawing may, for various reasons, go in and out of a drawer, but that isn't going to be its final location. The purpose of a drawing is to help assemble machine parts or to show the layout and construction details of some project or a building. It may be used in some manufacturing process. A number of duplicates may be made of your drawing by means of a photographic process. For these reasons you should be careful to have your drawing as clean as possible. A sloppy drawing, even though it may be correct in every detail, may be difficult to read. It also may cause errors in the object you are trying to build, in manufacturing or construction, and doesn't inspire confidence.

To keep your drawing clean, follow these few simple rules. After using them a few times, you'll find yourself following them automatically.

- Before you begin a drawing, take a clean cloth and wipe both sides of the T-square. Pencil smudges on the T-square can transfer to your drawing. Make sure the work surface of your drafting board or table is clean by wiping it with a cloth. Be sure to clean your instruments also.

- As you do your work, you may find it necessary to erase. After you erase, dust the erasure crumbs from the board and wipe the under side of the T-square. You may find the crumbs have become electrically charged, developing a tendency to cling to non-metallic surfaces. If you do not remove the crumbs you will find them deflecting your pencil, resulting in lines that aren't absolutely straight. They will also keep your T-square and triangles from resting flat on the paper on which you are doing your drawing.

- Never use a lead pencil softer than a 2H. Soft pencils will smudge your work. You will also find that the underside of your T-square will become coated with graphite, producing large smears as you move the T-square.

- Keep your hands clean. If not, you'll soon see the dirt on your hands being transferred to the drawing. You're going to find it impossible to keep your hands off the drawing since you will need to lean on it for support when using some tools. Hands off is ideal, but isn't an easy rule to follow. Some draftsmen use a sheet of paper to cover most of the drawing area, leaving exposed only that section of the drawing being worked on.

- When you go to lunch or when you finish your work for the day, don't leave your work exposed on the drawing board. To protect your drawing from dust or damage, be sure to cover it with a sheet of clean paper. Another reason for covering work is to keep it away from the eyes of unauthorized personnel. Drafting work is often regarded as confidential material.

- There is still one more reason for cleanliness. In many drafting departments a log or record is kept of the time required for the completion of a drawing or any section of it. Nonproductive drawing time is included in the total. You can be sure that the drafting supervisor has a very good idea of just how long a particular drawing should take. Draftsmen who have sloppy drawing habits and use an excessive amount of time doing their work become "job hoppers," moving from one drafting company to the next.

FASTENING THE DRAWING SHEET

To start actual drafting work, all you will need are a drawing board, a T-square, one or more triangles and a small roll of narrow masking tape. The first step is to mount the paper on the board. It may seem that fastening a sheet of paper to a drawing board should be an easy thing to do, yet unless you do this correctly the actual drawing may not be satisfactory. To fasten the drawing sheet to the board the correct way, follow these steps:

- Make sure your drawing board, T-square and triangles are free of all dust. You can do this with any section of clean scrap cloth. Just make sure the cloth has no buttons or fasteners on it that might scratch your equipment.
- Put the drawing sheet in position near the upper left-hand corner of the drawing board, just as shown in Fig. 2-1. The top edge of the drawing sheet should be parallel with the upper edge of the drawing board.
- Put the T-square into position on the board with the head of the T-square held firmly against the left edge of the board.
- Take a small section of masking tape and put it in the upper left-hand corner of the paper. The tape should extend slightly over the board but should not (at this time) make any contact with it.
- Now, using your left hand, slide the head of the T-square into position. Make the top edge of the paper parallel with the top edge of the T-square. If the paper is not parallel, adjust it until it is.
- Put the palm of your right hand on the paper and stroke the paper so that it adheres smoothly to the board. Work from the upper left-hand corner to the lower right-hand corner. Then place tape on the lower right-hand corner. Move back to the

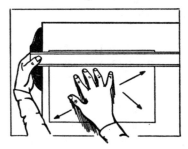

Fig. 2-1. This is how you can position your work sheet on the drawing board.

diagonally opposite corner — the upper left-hand corner — and press down on the tape you put there. Your sheet of paper is now being held in position by the two pieces of tape.

- Stroke the paper from the middle of the sheet to the upper right-hand corner and put another small piece of tape there.
- With your T-square, test the position of the paper again. Do this by checking the top edge of the paper.
- Stroke the paper from the middle of the sheet to the lower left-hand corner. Tape this corner into position and press firmly on the tape.
- At this time you should slide your T-square up and down the paper, with the head of the T-square held against the left-hand side of the board. The T-square should move smoothly and easily.

That sounds like a lot of work just to position a piece of paper. Perhaps. But it is less work than doing a drawing all over again because lines which are supposed to be horizontal are actually going up and down hill. It also takes more time to explain how to position the drawing paper than it does to do. With some practice and experience you will be able to position work sheets in just a few seconds.

Some draftsmen use pushpins for holding paper to the board, since they find them easier and faster to use than tape. The disadvantage is that the pins punch holes in the board. Further, the pins can interfere with the free movement of the T-square and triangles.

USING THE T-SQUARE

Now you are ready for your first drawing effort. It involves nothing more difficult than drawing some horizontal lines. Hold the pencil as shown in Fig. 2-2 and draw a line from the left to the right side. Do not draw from right to left. Move the T-square down and repeat. Lift the pencil at the end of the line, move the T-square down once more and then draw still another horizontal line. Try to make all your lines with uniform weight — that is, all the lines should look the same and should not vary in thickness. Always draw lines along the upper edge of the blade. Move the T-square up or down to position by sliding the head along the edge of the drawing board with your left hand.

Drawing Vertical Lines

You can draw vertical lines by using the T-square as a base for any one of the triangles. With the T-square in position, place one edge of the triangle against the blade and then draw your vertical line along the vertical edge of the triangle.

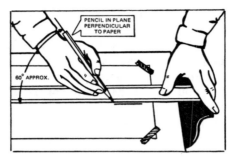

Fig. 2-2. When using a T-square, hold it firmly and flatly against the drawing board. Hold the pencil so it forms an angle of about 60° with the surface of the board.

You can use the same method for drawing lines at an angle of 30°, 45° and 60° to the horizontal T-square. All you need do is to turn the triangle over so that its slant side faces the left side of the board instead of the vertical edge.

Lines Of Reference

The bottom and top horizontal edges of your drawing paper can be considered as horizontal reference lines. Since any horizontal line you produce on the sheet is parallel to these edges, it can also be considered as a reference. Thus, a horizontal line on the paper forms an angle of 0° with the reference edge. If you draw a vertical line, it makes an angle of 90° with either the horizontal edge of the paper or with any horizontal line you want to draw.

Angles that are drawn on paper are usually considered with reference to the horizontal edge or some horizontal line, but can be referred to the vertical as well. The difference between the horizontal and vertical is 90°. Merely subtract the degrees of a horizontal angle from 90° to find the degree of the vertical angle corresponding to it, as shown in Fig. 2-3.

As an example, an angle of 30° from the horizontal will automatically be 60° from the vertical. Usually, when an angle is specified, it is with reference to the horizontal edge. If no edge is mentioned, you can assume that it is the horizontal that is intended. For example, if an angle of 37° is indicated, this means a slant line forming an angle of 37° with either the horizontal base edge of the drawing paper, or some line drawn parallel to it. When a vertical angle is intended, it is generally specified as such.

Now suppose you want to draw a line that forms an angle of 75° with the horizontal or 15° with the vertical. You can do this by using

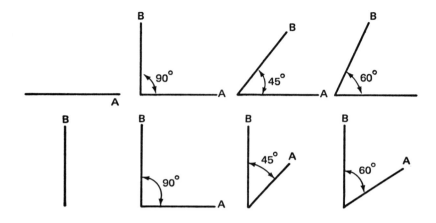

Fig. 2-3. Horizontal line A is the reference and is horizontal. The various drawings across the top show some of the possible angles formed by line B with respect to line A. Vertical line B is the reference. The drawings also show some of the possible angles that line A makes with respect to line B. In drawings, unless otherwise specified, a horizontal line is regarded as the reference.

two triangles — a 45° triangle and a 30° triangle. Rest the side of the 30° triangle against the T-square and then set the 45° triangle right above it. The 30° angle and the 45° angle add up to 75°. As a possible work exercise, you might try drawing horizontal lines, and then forming angles of 30°, 45°, 60° and 75°.

Parallel Vertical Or Inclined Lines

You can draw parallel vertical or inclined lines by sliding the triangle along the blade of the T-square as shown in Fig. 2-4. This technique calls for some practice since the head of the T-square isn't flush against the side of the drawing board. This means the T-square isn't being given edge support. You will need to press firmly on the blade to hold it in position.

You can draw lines parallel to any given line by placing the side of the triangle flush with the line and fitting the T-square or straightedge along its base. Then slide the triangle along the blade to any position you wish and draw a line.

You can draw a line perpendicular (at right angles) to another line by using the adjacent sides of a triangle. The adjacent sides are those which form a 90° angle. If one side is parallel with a line, the adjacent side will be perpendicular to it. Put the long side of your triangle against the straightedge and draw a line. Then slide the triangle along the straightedge and use the adjacent side to draw another line.

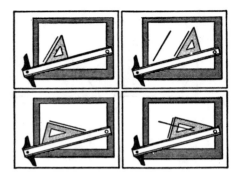

Fig. 2-4. You can draw inclined lines by tilting the T-square.

ANGLES AND CIRCLES

Just as an inch, a foot, and a yard are units of linear measure, degrees are units of circular measure. The curved line which forms a circle can be divided into 360 equal units, each of which is a degree. We can also divide the circle, as shown in Fig. 2-5, into four equal segments or quadrants. Since the circle is now cut into four parts, each of these quadrants is 90°. The horizontal line passing through the center of the circle is often referred to as the X axis. The vertical line, also passing through the center of the circle, is known as the Y axis. These two lines are perpendicular to each other. Since they form the outer edges of the quadrant, they enclose a total of 90°.

It makes no difference whether a circle is small or large, for no matter what its size may be, its circumference can always be divided into 360 equal parts or degrees. This means that the size of an angle does not depend on the lengths of its two arms, but only on the amount of opening between them, as illustrated in Fig. 2-6.

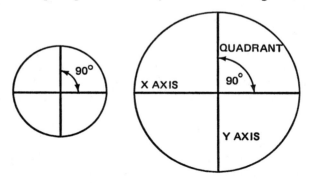

Fig. 2-5. All circles have a total angular measurement of 360°. The smaller circle at the left has a circumference of 360°. So does the larger circle at the right.

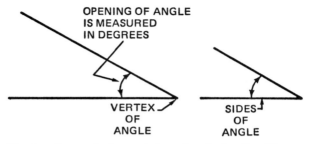

OPENING OF ANGLE
IS MEASURED
IN DEGREES

VERTEX
OF
ANGLE

SIDES
OF
ANGLE

Fig. 2-6. The size of an angle does not depend on the lengths of its arms, but only on the opening between them. The angles shown here are identical.

An angle can have any value from 0° to 360°. At 0° the two arms of the angle are superimposed on each other and similarly at 360°. In between you can have any angle of any size you wish. An angle that is less than 90° is called an acute angle. When the two arms are perpendicular to each other, the size of the angle is 90° and is known as a right angle. An angle greater than 90°, but less than 180°, is called an obtuse angle. These three basic types of angles are illustrated in Fig. 2-7.

Measuring An Angle

You can estimate an angle, roughly, just by looking at it. If the two arms of the angle are perpendicular, there is no need to measure, since the angle must obviously be 90°. You can also use your triangles to get some indication of approximate angle size. Thus, with the help of triangles, you can know if an angle is greater or smaller than 30°, 45° or 90°. However, for a more precise determination, you will need to use a protractor.

You can use your protractor either for measuring an existing angle, or for producing an angle of any size between 0° and 360°. Assume you wish to measure an angle. If you will examine the protractor you will see it has a tiny opening or some sort of mark at its center. Put this opening or mark at the point where the two arms of the angle meet. This juncture is called the vertex of the angle. With the protractor in this position, rotate it until the division marked zero on the outer edge of the protractor lies right along the horizontal axis of the angle. The other arm of the angle will now pass through some section of the protractor. All you need do is to read the number shown on the protractor and that will be the size of the opening in degrees.

Figure 2-8 shows a method for either measuring an angle or for drawing one. Assume you are using a circular protractor and wish to measure the size of angle AOB. Note that the center point, O, of the protractor now rests on the vertex of the angle, and that point a,

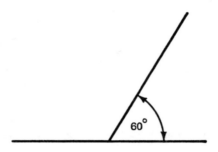

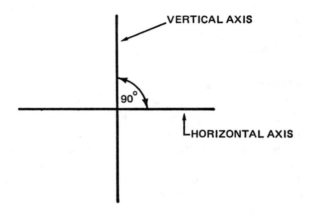

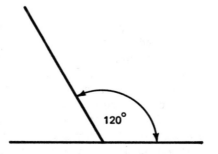

Fig. 2-7. At the top is an acute angle. In the center is a right angle. At the bottom is an obtuse angle.

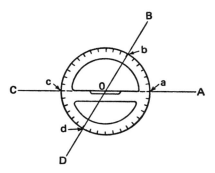

Fig. 2-8. Method of measuring an angle, or laying out an angle.

the zero point of the protractor, rests directly on horizontal axis AC. Line OB appears directly under the protractor. Examine point b on the protractor and you will be able to read the amount of the angle in degrees. Some protractors are even measured in half degrees, so you should be able to get a fairly accurate measurement.

Laying Out An Angle

To lay out an angle, follow the same procedure, but in reverse. First, draw a horizontal line using your T-square. Using your T-square and the vertical side of any triangle, erect a perpendicular line to point 0 as shown in Fig. 2-9. Select any point along the horizontal reference line as the possible vertex of the angle you want to draw. We call this point 0. Put the center point of the protractor right on point 0. Examine the circular edge of the protractor and select the size of the angle you want. Using any triangle draw a pencil line from any point, to the center point 0.

You will now have two lines. One of these will be the horizontal reference line, and another line, a slanting line, which forms an angle with the reference line. And, incidentally, you can make the arms of the angle just as short or as long as you want. Further, they need not be of the same size. None of these factors will change the size of the angle.

With the help of your protractor and your T-square, try drawing angles of various sizes. Also draw some random angles and measure them with your protractor.

Drawing Angles Without The Protractor

You can use your T-square and a pair of triangles, a 30°–60° triangle and a 45° triangle for drawing angles from 0° to 360° in steps of 15°. However, for drawing angles having intermediate values, you will need to use a protractor.

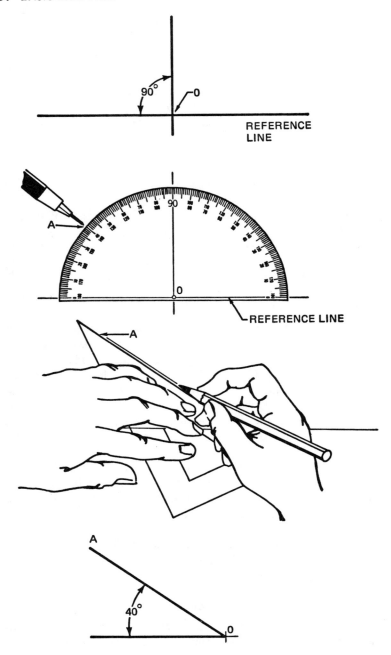

Fig. 2-9. Method for laying out an angle using a protractor and a triangle.

To produce angles with the T-square and triangles, start by drawing a horizontal reference line with the help of the T-square. Reset the triangle on the blade of the T-square and draw a vertical line to the reference line. Figure 2-10A shows how to draw a 45° angle using a 45° triangle.

With the help of two triangles you can draw a variety of angles. In Fig. 2-10B we have a 30°–60° triangle and a 45° triangle positioned as shown. With these two triangles we can erect a 30° angle, then another 30° angle, a 45° angle and then another 45° angle. The two angles at the center combine to produce a 75° angle (30° + 45° = 75°).

Figure 2-10C illustrates how to produce 15° angles while Fig. 2-10D indicates the variety of angles that can be obtained with these two triangles. Since a circle can be divided into four equal parts, or quadrants, you can repeat these angles to the left of the vertical axis, or below the horizontal axis.

With your T-square, 30°–60° triangle and 45° triangle, draw angles of 45°, 60°, 75° and 120°. After you finish, measure the angles with your protractor.

Drawing A Circle

Before drawing a circle there are a few points you should think about. The first is the location of the circle on your drawing paper. Still another is to decide on the center point of the circle, and, finally,

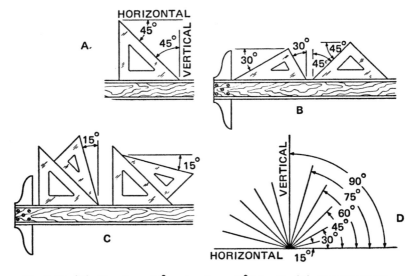

Fig. 2-10. (A). Drawing a 45° angle using a 45° triangle. (B). The two angles at the center combine to produce a 75° angle. (C). Here's how to produce 15° angles. (D). A variety of angles can be obtained with the two triangles.

the radius of the circle. A diameter of the circle is any straight line drawn from any point on the circumference, through the center point of the circle, to the opposite point on the circumference. A radius is always equal to one half of a diameter. The radius can be any straight line from the center point of the circle to any point on the circumference. This applies to any circle. All the radii of a circle are equal to each other, and all diameters of any one circle have the same dimensions.

To draw a circle, rest the needle point of the compass on the drawing sheet. Holding the compass as shown in Fig. 2-11, gently rotate the marking leg until you have formed a smooth, continuous circle. Ideally, the circle you draw should be a line of uniform thickness. It should not have any dark or light sections, nor should there be any breaks in it. At first you will find the needle point has a tendency to puncture the paper. After some practice you will be able to have this point indent the paper without piercing it, forming a tiny depression, but large enough to support the rotating needle point.

If you want to locate a circle at a particular area on the drawing paper, use your T-square to draw a small horizontal line. Then use the vertical edge of any triangle to draw the intersecting vertical line. The point at which the two lines meet will be the center point of the circle, and is the point at which you will rest the needle point of the compass.

To draw a circle having the radius you want, put the compass alongside a scale and adjust the compass center wheel. This will supply the radius of the circle. After you have set the compass to the radius you want, be careful not to disturb the center wheel of the compass. If you want extreme accuracy when drawing circles, or if you need to draw very small or very large circles, it is best to have

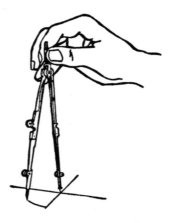

Fig. 2-11. How to use the compass to draw small circles.

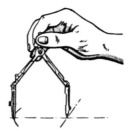

Fig. 2-12. With hinged legs, a compass can be used to produce more accurate circles.

both legs of the compass so they are perpendicular to the paper. A compass may have hinged legs. These can be adjusted so they have the correct relationship to the worksheet, as shown in Fig. 2-12.

USING A FRENCH CURVE

On your drawing you may have a number of points, but these points, when connected, will not form a straight line. At the beginning you may find it helpful to sketch in very lightly a series of connecting lines between the points to give you some idea as to the shape of the curve. Naturally, the closer the points and the more points there are, the easier this will be to do.

Select a French curve and place it alongside the points on the drawing. The idea is to find a curve that will lie along as many of the points as possible. This cannot always be done. You may have to select a few points at a time, points that are adjacent to each other. To draw a complete curve, you may need to draw a succession of small curved lines. At no time should any of these lines form a sharp angle with each other. Rather, the line should be smooth and continuous.

USING DIVIDERS

If you want to make a measurement with the dividers, put the points of the dividers on the line to be measured. Without changing the position of the legs of the dividers in any way, transfer the instrument to a scale and read off the measurement directly from the scale.

Figure 2-13 shows how you can divide a line into equal parts. Decide on the length of line you want and set the dividers to half this distance by putting the points of the divider on a scale. Move the dividers over to the line without changing the legs of the dividers in any way. Start at the beginning of the line with one divider point and let the other divider point rest some distance away on the line. Press down lightly to make an impression, but do not punch through

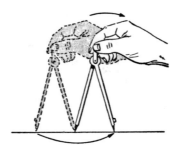

Fig. 2-13. How to divide a line into equal parts with a divider.

the paper. Lift the dividers and repeat the process, using the impression as your starting point. You can help identify the impression points by a pencil.

Figure 2-14 shows how you can divide a line into three equal parts. What you will follow will be a trial and error process. Adjust the dividers until they are approximately one-third the length of the line. Put one needle point at the start of the line and then see if you can take three equal steps to the end of the line. It is very unlikely you will succeed with the first try. You will either finish beyond the line or before its end. Adjust the dividers until you start at the beginning of the line and terminate at its end. You can then use the dividers and a scale to decide the actual measurement of one-third of the distance. Also, you may find it helpful to put a tiny pencil dot at each divider swing point on the line.

DRAFTING MEDIA

The various materials on which you will do your work are known as drafting media. Paper is one form of drafting media. Cloth is another. Tracing paper is still another and so is film. Within each category, such as paper or cloth, there are many variations with respect to overall size, weight, thickness, surface finish, color, etc.

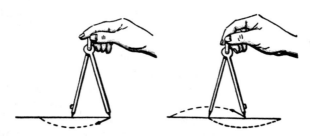

Fig. 2-14. How to divide a line into three equal parts.

Detail Paper

Detail paper is a heavy weight, opaque (non-transparent) paper which is usually buff or neutral green in color. It takes pencil well, but pencil lines on it are difficult to erase because a hard, sharp pencil makes a physical impression on the paper's heavy texture. To become more familiar with detail paper, try drawing a variety of lines on it, using different pencils. Your purpose will be to draw a clear, distinct line, uniform throughout its entire length which will not smear and which will make the least impression. After you draw a line, examine the erased section. Has the pencil made a groove in the paper? Can you draw any new lines over or through the erased area? If the paper has become grooved, you will find it collecting dirt, gradually showing the erased line more and more distinctly. Drawings are generally made on detail paper with the idea that these are to be used for tracing.

Tracing Cloth

Tracing cloth is a finely woven fabric. It is coated heavily with a starch-like compound and is then put through machine rollers to produce a glossy surface on one side and a dull finish on the other. At one time the glossy side was intended to be used as the working surface. Drawing on the glossy side results in ink lines which have sharp edges. It is also much easier to erase on the glossy side.

However, the dull side is now used as the working surface because it takes ink much more readily. The dull side can also take pencil lines, an advantage when it is necessary to indicate corrections or changes.

Tracing cloth is much more expensive than paper, but it does have a large number of advantages in its favor. It is transparent, strong, and can be used whenever a permanent drawing or tracing is essential. It can be stored for a long time without being affected. It can be erased, when the erasure is done correctly, without damage to its working surface.

Linen

You will sometimes hear reference to linen or to linen cloth as a drawing material. So-called linen cloth is really cotton. The surface of the cotton is treated with starch, and as a result the surface can take either pencil or ink. However, the cloth remains transparent and you can use it for tracing drawings placed beneath it.

Tracing Paper

Another type of paper you will be using is called tracing paper or tracing vellum. The paper is white or it may be slightly tinted. Tracing paper, as its name implies, was originally designed especially for tracing drawings. However, tracing a drawing takes time and can lead to errors. For these reasons tracing paper is now often used as the master drawing from which reproductions are made. Because of the use to which tracing paper is now being put, it is made as a much stronger more durable sheet than before. It can withstand repeated erasures, can tolerate frequent handling and has a surface on which pencil can be used.

You will find considerable variations in types of tracing papers. Before starting to work using such paper, try to learn what you can about its properties. Using a scrap sample, try it for pencil and ink. Try erasing it. Will it take erasing without tearing? Can you draw over an erased area without blurring or smearing lines? And, if you do plan to use it for tracing, does it have adequate transparency? Some tracing papers are more transparent than others. You must also determine whether you are going to trace simply by putting the tracing paper directly over the original drawing, or whether you are going to use a light box, such as the one described earlier in Chapter 1. With the help of a light box, the paper need not be as transparent, and it can be somewhat thicker and sturdier. New paper manufacturing techniques are constantly being developed with a view to making papers stronger, more transparent and easier to erase.

Pencil Versus Ink Drawings

At one time all finished drawings in a drafting room were done in ink. There were good reasons for this. Ink is much more permanent than pencil. When ink is dry, it does not smear. Ink has greater contrast than pencil and, for that reason, shop workers using drawings in manufacturing prefer ink drawings because of their greater legibility. Corrections in pencil drawings can sometimes lead to confusion if the pencil lines have formed grooves in the paper.

Today, however, most finished drawings are being done in pencil. The advantage of pencil over ink are several. Pencil erasures are easier to do. There is no danger of ink spilling. The time required for the care of a pen is eliminated. The disadvantages of pencil that we have mentioned have largely been eliminated through various methods now available for making duplicates of prints. With a machine such as Xerox, or similar equipment, it is now possible to make duplicates which not only do not smear, but which have the effect of emphasizing and strengthening the line structure.

Phototracing

Still another duplicating method is called phototracing. A phototracing is a photographic reproduction of a drawing on sensitized tracing cloth. The original drawing is photographed, and, as in usual photography, a negative of the drawing is obtained. The negative is then placed in close, firm contact with the sensitized tracing cloth. An intense light is beamed down on the negative and, as a result, a duplicate of the original drawing is obtained on the cloth. Since the light does not wear out the negative, as many duplicates of the original can be made as required. In a factory, it is not uncommon for a dozen or more duplicate prints to be made for various manufacturing departments.

The advantage of a phototracing, and also in the case of Xerox or other duplicating machines, is that there is no wear and tear on the original drawing. For this reason the problem of paper durability is eliminated. Original drawings can be made on less expensive paper. After the phototracings or Xerox copies have been made and distributed to the persons requesting them, the original drawing is put into a file vault where it is available for additional phototracings when required.

There are other advantages in using phototracings. Even if the original is a pencil drawing the phototracing looks like ink. For this reason, the phototracing will have better contrast and legibility than the original. And, unlike a pencil drawing, a phototracing will not smear.

Blueprints

The blueprint method for making duplicates of original drawings is one of the oldest. The technique is essentially the same as in phototracing and is a photographic reproduction of a drawing. The disadvantage of blueprints is that they work best if the original drawing is done in ink on cloth or vellum. A blueprint can be made of a drawing in pencil provided the pencil lines are sufficiently black.

One of the advantages of a blueprint is that it produces a reverse drawing. This means that all drawn lines will appear white instead of black, and the white area of the drawing will appear as a dark blue in the blueprint. This heightened contrast often makes the blueprint much easier to read than the original drawing. The readability of a blueprint depends on the line structure of the original drawing (that is, the sharpness of the lines), and the technique of the operator in making the blueprint.

Photostats

The various methods of reproducing original drawings described so far result in duplicates which have the same size as the original. In some instances, it may be desirable to have copy prints (duplicates of the original drawing) in a larger or a smaller size. This can be done by having a photostat made, more often referred to as a stat. A stat is a photographic reproduction of a drawing and is made to appear directly on the surface of prepared paper with the image in correct position and not reversed as in a negative.

Stats are available reduced, same size or larger than the original. But when the size changes, remember that it changes in two directions; that is, the width is reduced proportional to the depth. A stat that is "half off" means a 50 percent reduction in size. Thus, if the original is 4 x 8, the stat will be 2 x 4. The work "up" is sometimes used to indicate enlargement. Half up means that the stat is to be 50 percent larger in its dimensions than the original. The word "down" is used in the same context; 25 percent down means a stat that is one-fourth reduced in size from the original.

The original drawing may be on either transparent or opaque paper or cloth. When produced, the stat consists of white lines on a black background. A reverse can also be obtained, producing a result similar to a blueprint, but one in which the background is black instead of blue. The drawn lines, however, will be white.

Possibly the greatest advantage of using stats is the ability to get an increase or reduction in size. Most other reproducing processes do not have this advantage.

Microfilm

Storage of drawings may become a serious problem since thousands may be required in a particular manufacturing process. This means a large amount of floor space must be given over to files. Further, the files may pose a security problem, since drawings do represent a manufacturer's confidential information. The problem can be solved through the use of microfilm.

Microfilm is available in four sizes: 16mm, 35mm, 70mm, and 105mm. Compared with the 16mm microfilm, the 35mm film is larger; therefore, it provides a better reproduction of drawings. The number of drawings that can be photographed depends on the size of the drawings and the length of the film. To be able to read the film, a projector or a reader can be used. The advantage of microfilm is that several reels can duplicate thousands of drawings while requiring comparatively little room.

B-W Prints

As you can see, there are many methods for reproducing prints. Still another is called the B-W process. Using this technique, the tracing or the drawing and a sheet of sensitized paper are exposed to light. The sensitized paper is then fed through rollers which put a thin film of developer on it. The completed print consists of dark lines, but these lines can be black, brown or maroon, depending on the kind of developer that is used. The background is white. A print produced by this, or some similar method, having brown lines is sometimes referred to as a "brownline print" or simply a "brownline."

Ammonia Process Prints

The reproduction of a print can involve either a dry, a wet or a vapor process. The ammonia process uses ammonia vapor as the developing agent. As in the case of most reproducers, the drawing or tracing and the sensitized paper are placed in close contact and are then exposed to strong light. The sensitized paper is then placed in a container where it is exposed to ammonia fumes.

Any "wet" process of reproducing tracings or drawings does have problems. A process of this kind is usually much slower than a dry process. Further, a wet process sometimes distorts the paper, producing a reproduction that isn't entirely flat.

THINGS TO DO

- When you next visit an "art supply store" or a store that sells drafting supplies, examine some of the various papers they have available for drawing and tracing.
- Mount drawing paper of any convenient size on your drawing board. After doing so, check the paper to make sure its horizontal edges are parallel to the horizontal edges of the board.
- Using the T-square, draw a number of parallel lines.
- Using a triangle, draw lines perpendicular to the horizontal lines.
- If necessary, mount another sheet of drawing paper. Draw several horizontal lines, and then, with the help of triangles, draw angles of 30°, 45°, 60° and 75°.
- With the help of the compass and a scale, draw a circle having a radius of 1 1/2 inches.
- Draw a circle having a radius of 2 inches.
- Draw a line of random length. Using the dividers, mark off the line into two equal parts and then into three equal parts.
- Draw angles of various sizes and measure them with your protractor.

SUMMARY

- Vertical lines can be drawn by using the T-square as a base and the vertical edge of any triangle.
- The bottom and top edges of a work sheet can be considered as horizontal reference lines.
- Inclined lines can be drawn by using the slant edge of a triangle.

- All circles are 360°. They can be divided into four equal quadrants of 90°.
- Angles can have any value from 0° to 360°. An angle that is less than 90° is an acute angle. An angle greater than 90° but less than 180° is an obtuse angle. An angle of 90° is a right angle. Angles can be measured with a protractor.
- A diameter of a circle is any straight line drawn through the center point of the circle and touching the circumference in two places. A radius is equal to one-half the length of a diameter. A radius is the shortest straight line that can be drawn from the center point of a circle to the circumference.
- Curves can be drawn with the help of a French curve.
- Detail paper is heavy weight, nontransparent paper, usually buff colored or neutral green.
- Tracing cloth is finely woven fabric. The dull side is generally used as the working surface.
- Linen cloth is cotton treated with starch. The surface can take either pencil or ink.
- Tracing paper, also called tracing vellum, is used for tracing drawings.
- A phototracing is a photographic reproduction of a drawing on sensitized tracing cloth.
- A blueprint is a method for making duplicates of original drawings. In a blueprint the drawn lines are white; the background is blue.
- A photostat (or stat) is a photographic reproduction of an original drawing. It can be made same size, reduced or enlarged.
- Microfilm is photographic reproduction of drawings and is available in four sizes: 16mm, 35mm, 70mm and 105mm.
- In a B–W print the drawing is transferred to a sheet of sensitized paper. When the reproduced lines of the drawing appear brown the print is called a brownline print or a brownline.
- Ammonia process is a vapor process for the reproduction of a print.

Chapter 3
The Meaning of Lines

A drawing consists of numbers, words and lines. Although we usually think of numbers and words only as having meaning, lines also have something to say.

The purpose of a drawing is to convey precise, specific data, not generalizations. To an experienced draftsman, each line explains something about the object that has been drawn. A break or an interruption in a line, a dashed line, a dark line or one that isn't so dark all have meaning.

In drafting the meaning of a line is sometimes referred to as a line convention (Fig. 3-1). In other words, lines form a sort of vocabulary. If you understand the meaning of this vocabulary, then you can look at any drawing and understand exactly what the draftsman intended to say. Whether or not a draftsman conveys his full meaning depends on the expertise of the draftsman—his ability to include all necessary lines, or to omit those which do not add to the meaning or which tend to obscure it. It is possible, in a drawing, to have more lines than are needed, just as in conversation more words may be used than required. Unlike conversation, however, in a drawing it is always best to have the least amount of lines which will supply exactly the information that is wanted.

TYPES OF LINES

A variety of different kinds of lines are used in drafting, with each of these lines having a special meaning. Some of these lines have different weights; that is, some are thicker or stronger lines than others. Some are continuous; others are dashed. Some are straight; one is wavy. A collection of these lines could be called an alphabet and some of them are shown in Fig. 3-1.

Center Lines

Examine and you will see that a *center line* is made of long and short dashes (Fig. 3-1). These dashes are evenly spaced and each center line starts and ends with a long dash. They are sometimes

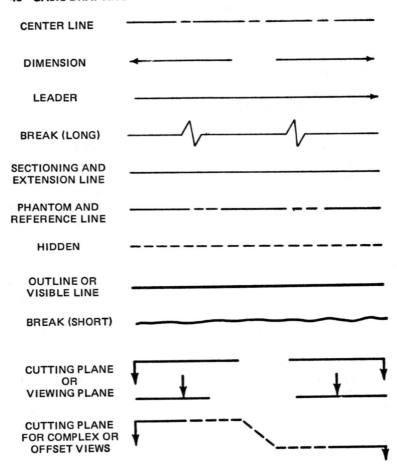

Fig. 3-1. Lines used in drafting. While there are other lines for specialized applications, such as marine, plumbing and electrical drafting, those shown here are basic to all types of drafting.

called primary center lines and are the lines on which the other details of a drawing could be based.

Any number of center lines may be required on a drawing (Fig. 3-2). All those other than the primary center line are called secondary. Secondary center lines, for example, are used with drawings of holes. Draw center lines lightly (Fig. 3-3).

Both primary and secondary center lines are also used to indicate the center of a symmetrical object (Fig. 3-4). This means that the parts of the object on either side of the center line are mirror images of each other. A center line may be horizontal, or vertical, and a drawing may

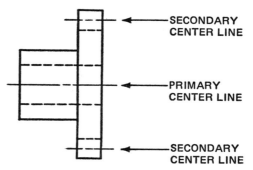

Fig. 3-2. The dashed line in the middle is the primary center line. The dashed lines near the top and the bottom are secondary center lines. Note that the center lines extend beyond the outline of the drawing and consist of long and short dashes. The equally spaced dashed lines shown in this drawing are not center lines.

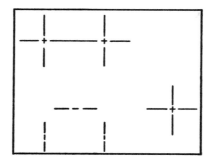

Fig. 3-3. Draw center lines lightly.

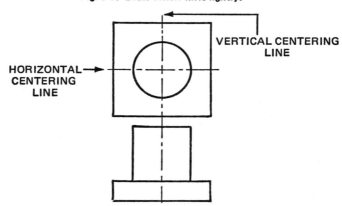

Fig. 3-4. Horizontal and vertical centering lines. These lines are always perpendicular to each other. You can use any number of such lines as needed: a single vertical and a single horizontal; a single vertical and several horizontal; several vertical and a single horizontal. Centering lines always use long and short dashes. In this drawing the vertical centering line continues through the two views of the same object.

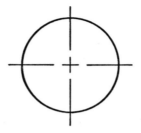

Fig. 3-5. A circle requires two center lines: one vertical, the other horizontal.

require the use of both. The drawing of a circle, for example, means two center lines are required (Fig. 3-5). The intersection of the center lines is the center point of the circle. Center line intersections are also used to indicate the radius point of an arc (Fig. 3-6).

Sometimes it may be necessary to draw a very short center line. In such instances a single short dash is permissible provided there is no possibility of confusing the center line with other lines.

Concentric circles consist of one circle (or more) drawn inside another (Fig. 3-7). Both circles use the same pair of center lines. Figure 3-8 shows some of the other ways in which center lines can be used.

Visible lines

Visible lines, as shown in Fig. 3-9, are solid, thick lines, thick, that is, in comparison with center lines. Visible lines define the outline of the object being drawn and for that reason are sometimes called "outlines."

Figure 3-9 is a drawing of two different views of the same object. In the drawing at the left you are facing the object, while in the drawing at the right the object has been rotated 90°, and so you have been presented with a side view.

Now compare the visible lines or "outlines" with the center lines and note the difference. The visible lines are darker and are unbroken. Note that the center lines cut across and extend beyond the visible

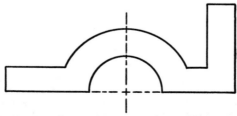

Fig. 3-6. Center lines are used for arcs since they are part of the circumference of a circle.

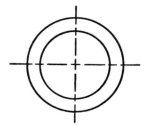

Fig. 3-7. Concentric circles share center lines.

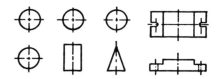

Fig. 3-8. Examples of the use of center lines.

lines. The drawing at the left in Fig. 3-9 has a pair of center lines, while that at the right has just a single center line. The center line at the right may be considered an extension of the horizontal center line at the left. You can check this by putting the edge of a triangle or T-square on the left center line and noting its relationship to the center line at the right. They both seem part of the same line.

Hidden Lines

An object that may seem solid to anyone else may not appear so to a draftsman. He may be aware that the object has one or more holes drilled in it. If he is to represent the object as it really is, he must have some way of indicating what is hidden by the surface.

A hole, for example, may be illustrated by means of hidden lines, as shown in Fig. 3-10. You can represent hidden lines by using short dashes of equal lengths and equally spaced.

If you will examine the drawing at the left in Fig. 3-10, you will see it consists of a square with a circle. By itself this drawing does not con-

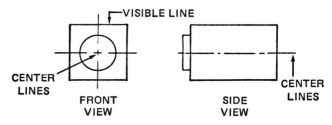

Fig. 3-9. Visible lines are thicker than the center lines.

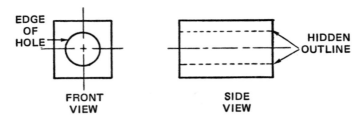

Fig. 3-10. Technique for showing hidden lines.

vey all the information. However, if you will also look at the drawing at the right, you will see it has a pair of dashed lines. These dashed lines tell us that this object has a cylindrical hole cut in it. The hole is cut completely through the object, starting at the front and ending at the back. We know this since the hidden lines extend from front to back. If they extend only halfway, then we would know that the hole started at the front and stopped in the center of the object.

Hidden lines start in contact with a visible line. There should be no space between the visible line and the hidden line. Be careful to note the difference between the construction of center lines and hidden lines, and where they stop and end.

In Fig. 3-10, in the front view, the edge of the cylindrical hole is identified by "edge of hole." Normally, these words wouldn't appear on a drawing, but are used here to call your attention to the meaning of the circle. The hole in Fig. 3-10 is sometimes called a "through hole" since it extends completely through the object.

Now consider why we show hidden lines in the drawing at the right and none in the drawing at the left. With the front view we can actually see the hole. We look right at it. But with the side view we can no longer see the hole or any part of it. And so we must represent the hole by dashed lines, a hidden outline.

As a practice exercise draw several circles, each having a different diameter, and then draw vertical and horizontal centering lines. Also draw a pair of concentric circles and include the centering lines. Copy the lines shown in Fig. 3-11 and try to memorize them.

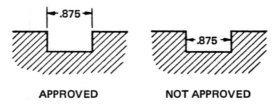

Fig. 3-11. Extension lines are used for showing sizes.

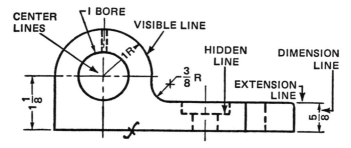

Fig. 3-12. Extension lines aren't always required, but are used as needed.

Extension Lines

To manufacture an object we must know how big it is. We must know not only the overall dimensions but also the sizes of holes, circles and parts of circles. It is important for dimensions to be out in the open where they can be clearly seen and read. Some drawings, because of the nature of the object to be manufactured, require a large number of dimensions. These must be placed so they do not interfere with the line structure of the object in the drawing. To be able to do this, we use *extension lines.* These are straight, thin lines which lead to some part of the object.

How do you draw an extension line? This is shown in the drawing at the left in Fig. 3-11. The line — actually a pair of lines — is a thin, unbroken line starting about 1/16 of an inch away from the visible line. Fig. 3-11 also shows how extension lines are used to move the dimensions away from the body of the drawing. The drawing at the left in Fig. 3-11 indicates the proper way of dimensioning. You will sometimes find the dimension technique shown at the right used in some drawings, but it is not regarded as desirable. When drawing extension lines, make sure they do not touch the visible lines since extension lines are *not* part of the object.

Extension lines aren't always required, for there will be times when you will be able to put dimensions on a drawing without their help. In Fig. 3-12 for example, we have extension lines at the left in the drawing and also at the right. But the radius of the fillet shown as 3/8 near the center part of the drawing does not require an extension line.

Dimension Lines

As its name implies, the purpose of a *dimension line* is to call attention to a dimension. The dimension line ends in an arrowhead,

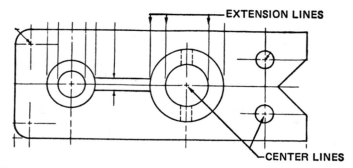

Fig. 3-13. Sometimes a drawing will require a large number of dimension lines. Dimensions, not yet drawn, will be made between pairs of extension lines.

or a pair of arrowheads. The dimension line is usually broken; that is, it is not a continuous line to permit the insertion of the dimension. Dimension lines are used in conjunction with extension lines.

Sometimes an extensive amount of dimensioning may be required for a drawing, such as the one shown in Fig. 3-13. If dimension lines must cross each other, they should be broken. If you will examine the two drawings shown at the top of Fig. 3-14, you will see why the one at the left is preferred. There is less chance of confusion. When dimensioning a drawing, you will find that a few minutes spent in planning extension lines will be worthwhile.

Draw extension lines and dimension lines thinner than visible lines. One of the problems in a drawing may be the need for a large number

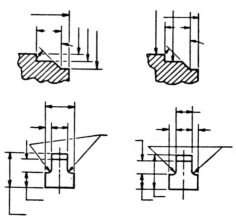

Fig. 3-14. Approved method of drawing extension lines (upper left) and undesirable method (upper right). Extension lines may be broken when they cross each other. The preferred method is at the lower left. Technique at the lower right isn't regarded as desirable.

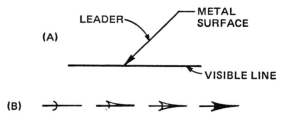

Fig. 3-15. (A). A leader is used to call attention to some property of the object. (B). Steps in drawing arrowheads.

of dimension lines. To avoid confusion it is preferable to have dimensions spaced away from each other by at least 1/4 inch. Remember that a drawing is not an end in itself, but is a set of instructions. The important fact is not that you completed the drawing, but that someone else must read it. And so you must make your drawing as legible as possible. One technique for doing this is to avoid overcrowding dimensions, and giving some thought to their spacing.

Leaders

A *leader*, as shown in Fig. 3-15A, is a line with an arrowhead. The arrowhead touches the visible line. Leaders are used to call attention to some property of the object being manufactured. The leader may indicate the type of surface, the kind of metal being used, or it may have references of particular importance. Note that the leader line is drawn at an angle. The reason for doing this is to emphasize the fact that the leader is not a construction line.

The leader should end in a line which is drawn parallel to construction lines (visible outlines). Observe the positioning of the lettering following this parallel line. If the parallel line were continued, it would intersect the center of the first letter. The lettering should also be parallel to the construction lines.

Drawing Arrowheads

Arrowheads are often used in drafting, especially in drawings that require dimensioning or which need the use of leaders. Although it may seem like a minor detail, arrowheads are important. If drawn in a sloppy or incorrect manner, they can spoil the appearance of what would otherwise be a professional job. The size of arrowheads can vary from one drawing to the next, but all those used on a single drawing should be the same size.

Figure 3-15B shows the correct way to draw arrowheads. First, draw a short straight line as shown. The length of this line will determine the size of the arrowhead. This line should intersect a small arc placed to the left of center. Connect the ends of the arc to the right side of the line. Finally, fill in the arrowhead.

You can save time in drawing arrowheads by using a template. This has the further advantage in insuring that all arrowheads will have the same shape.

Break Lines

Sometimes the object being drawn is a continuous length of identical material. For example, a pipe that is quite long could be cut into a number of identical sections. Nothing would be achieved by drawing the entire pipe. Instead, for objects that are uniform, such as pipe, *break lines* can be used to save drawing space. This drawing technique can be applied to any object that is the same throughout its length. Thus, metal rods, metal tubing, metal bars and wooden beams may all be individually uniform.

Figure 3-16 shows various methods that are used to indicate breaks in construction lines. Starting with the drawing at the top, the shaded area indicates that the metal rod is solid. The drawing of the metal tube shows that it has an interior that is hollow.

Of course we don't depend only on the break lines to tell us that an object is solid or hollow. If you will examine the drawing of the metal tube, you will see that we have also used hidden lines. These hidden lines indicate that we have a tube with a wall of a certain thickness. Also, compare the end views of the metal rod and the metal tube. Note that the end drawing of the metal rod is a circle. If you were to examine the end of the metal rod, that is what you would see — an area having a circle as its outline. The end view of

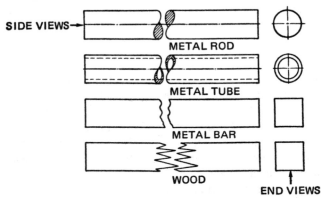

Fig. 3-16. Techniques for showing breaks in construction lines.

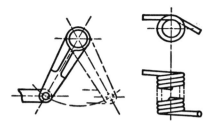

Fig. 3-17. Phantom lines show different positions that can be taken by an object.

the tube, however, consists of a pair of concentric circles. The outer circle represents the outer circular surface of the tube. The inner circle indicates the inner circular surface. The area of the inner circle in the end view is the hollowed out interior of the metal tube.

Now consider the metal bar and the block of wood (Fig. 3-16). Both of these are solid objects. If you will look at the end views of each, you will see they are both represented by squares. Note that the two end views are identical and that they tell us that both objects are solids.

Phantom Lines

Sometimes a drawing may show the alternate positions an object may have when it is completed. The alternate position is indicated through the use of phantom lines, as shown in Fig. 3-17. The *phantom lines* are lighter than the visible lines and are represented by long and very short dash lines. They are similar in construction to center lines.

A straight or curved line can be used to indicate the motion of the object. In the drawing at the left in Fig. 3-17, the object is pivoted at the top and swings in the form of an arc. This is shown by the curved line near the bottom of the drawing.

The drawing at the right in Fig. 3-17 is that of a spring. The parallel dashed lines at both sides show the possible up and down movement of the spring. The drawing at the top right is a view looking down on the spring. Note that three types of lines were used in making this small drawing. We have the usual center lines, both primary and secondary, the visible lines, or construction lines, and the phantom lines.

Cutting Plane Lines

In drafting it is often customary to look at an object from three different views: front, top and side. Other views are possible, and so

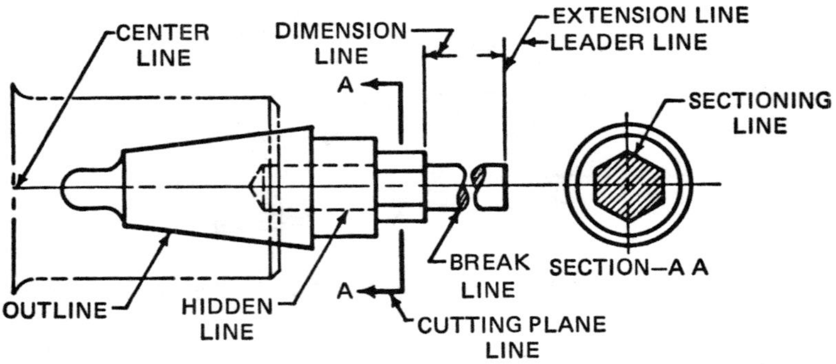

Fig. 3-18. This drawing illustrates the use of lines described earlier.

you will have views with the object positioned at some other angle. These views are often not enough. While they show the outside of the object, we still know nothing of what is covered by the surface. If we are not interested in what is inside the object, then outside views alone are satisfactory. However, we may want to know what is inside; that is, we may want to see a cross section of the object. The action of a cross section is just as though you had taken a knife or saw and had cut through the object.

Figure 3-18 shows an object as it might appear on your drawing board. Note the positioning of the cutting plane line in the drawing at the left. This indicates that the engineer or designer of the object wants to show a cross section of the object at that particular point.

Examine the method of drawing the *cutting plane line.* It is shown by lines above and below the object. Each cutting plane line consists of a long line and a short line at right angles to each other. The short line ends in an arrowhead. You can use letters to help identify the cutting plane line. In Fig. 3-18 the cutting plane line ends in the letter A.

The drawing at the right is the section produced by the cutting plane line. This section is the result of cutting through the object along the cutting plane line of the object at the left. After the cut, the object is rotated 90° so that the cut area faces us directly. The drawing at the right shows a side view of the object but with part of the object cut away. In addition to the section view, there might also be an end view.

Sectioning Lines

Note the use of *sectioning lines* to indicate the exposed surface in the drawing at the right in Fig. 3-18. These lines are thin and are drawn parallel to each other, but they may vary, depending on the kind of material shown.

Figure 3-18 is interesting since it illustrates all the lines that have been described. The visible lines are identified by the word "outline." Compare the weight of the visible lines with the other lines. Note also that the cutting plane line is quite heavy, but does not touch the visible lines.

LINE WEIGHT

The thickness of a line is often referred to as its weight. A heavy line is one that is wide; a light line is one that is fine. When doing pencil drawings you will find that it is not as easy to control the weight of the lines as in doing inked drawings. In inked drawings you can control line width very easily by adjusting the amount of opening of the ruling pen.

Lines in a drawing can be grouped according to function. Thus, some lines are intended for dimensioning, others as phantom lines, and others as visible lines. All lines in a particular category must have the same weight. This means you cannot have narrow lines and thick lines for dimensioning. Once you decide on the line weight for a particular category of line, use it throughout the drawing.

There are three basic weights of lines: thin, medium and thick. Thin lines include center lines, dimension lines, leaders, long break lines, extension lines and sectioning lines.

Medium weight lines include phantom lines, reference and hidden lines. Thick lines are used for outline or visible lines, short break lines, cutting plane lines (also known as viewing plane lines) and cutting plane lines for offset views. An offset view means that the cutting plane does not go straight through the object, but "offsets" or follows another path so as to supply another view.

LINE SUMMARY

The drawing shown earlier in Fig. 3-1 is a summary of all the lines that have been described. These are all basic lines and are used regularly. It does not mean that a single drawing will contain all of them, but you will be using them all as you do more drawings. A good way to remember them is to use a ruling pen and to draw them on scrap paper. But as you work with these lines, it is important to remember that you will be working with lines having three different weights. This means you will have to adjust your ruling pen for each weight of line.

GOOD WORK HABITS

To avoid frequent resetting of the ruling pen, it is good practice to ink in all lines of one weight and then adjust the drawing pen to produce

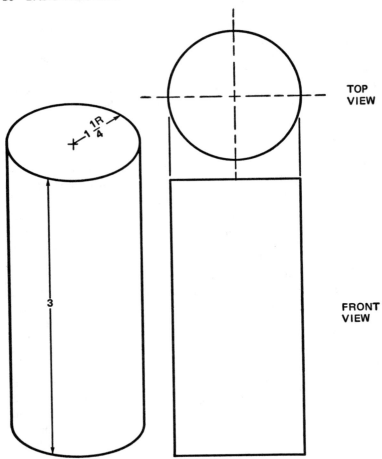

Fig. 3-19. A cylinder with top and side views.

lines of a different weight. Another reason for doing this is to make sure that all lines of one weight have the same thickness. Follow this procedure if you must also use a bow pen. To make sure that a line will have the right thickness, always draw a line on a bit of scrap paper before working on a drawing. And, if you are ready to ink in a drawing, you'll find it helpful to have a bit of scrap alongside to "start" the ink flowing out of the pen. If not, you may find your inked line starting a fraction of an inch after you have touched the ruling pen to the work paper. This means you will need to go back and fill in that part of the line not covered by ink. It isn't as easy to do this as you might imagine, for the two lines may not join perfectly, or the point of joining may look unusually thick.

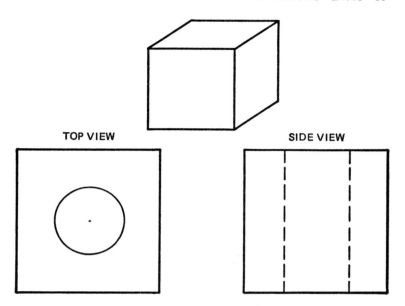

Fig 3-20. A cube with top and side views.

One of the problems you will face when working with ink for the first time is that the ink will seem to want to run beneath the guide, whether that guide is a triangle or a straightedge. You can avoid this by attaching a few small coins beneath the guide so it is above the surface of the paper. As you gain experience you will find this is no longer necessary. Ink has a tendency to flow under the guide when the pen is held so its top leans outward. If, on the contrary, you hold the pen so it leans inward too much, the result will be an irregularly shaped line. It takes some practice to avoid these two extremes.

After you finish your work, remove any excess ink from the pen with a cloth. You can remove dried ink by washing the pen in a weak solution of ammonia and water. Always be sure to put your instruments away clean so they will be ready for the next job.

PRACTICE EXERCISES

- Make a drawing of a cylinder, 3 inches long and having a radius of 1 1/4 inches, as shown in Fig. 3-19. Make the drawing in pencil. Note that the front view in this case looks like a rectangle, and the top view is a circle. Be sure to include the center lines.
- Dimension the drawing you have just made.
- Draw a cube with every edge having a dimension of 3 inches (Fig. 3-20). Draw the front and side views. Imagine that a

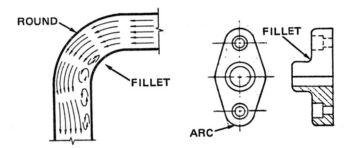

Fig. 3-21. The interior curved angle is a fillet. The outside curved angle is a round. These are often associated, but they can exist separately.

hole is drilled completely through the cube and that this through hole has a diameter of 1 inch. Show how this hole would appear in the front and side views. In the side view the hole would be represented by hidden lines.

- Practice drawing center lines, dimension lines, leaders, break lines, phantom lines and hidden lines.
- Draw a circle having a diameter of 1/2 inch, and dimension it for a tolerance of .001 inch.
- Draw a rectangle 1 inch long and 1/2 inch wide and dimension it.
- Try drawing objects with interior carved angles, or fillets, and outside curved angles, or rounds (Fig. 3-21).

Now that you have had your introduction to drafting, you can see it isn't difficult if you take it one step at a time. You are also going to start working with ink as well as pencil, and there are some other instruments in your drafting set you must learn about.

PENCIL WORK

Assume that the job you have in front of you is a drawing that is to be done in pencil. But whether the lines are to be done in pencil or in ink, they must follow a certain procedure.

Drawing requires a certain amount of planning and so you may find it helpful to organize your thinking. Make some freehand sketches on scrap paper of the various views you expect to show. Do this before you begin the actual drawing so that you will have some idea of your objective.

After you make your freehand sketches, cut them out and position them on a sheet of paper in the correct order and approximate location Allow space for any lettering you may need to do. Also

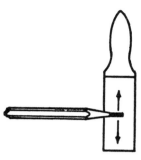

Fig. 3-22. Technique for shaping a pencil to a point.

remember you may need room for other information on the drawing, such as dimensions.

Finally, the freehand sketches you make should have approximately the same size as your final drawing. If you do this you will not be embarrassed, as you complete a drawing, to find you don't have enough room for everything that must appear on the drawing. There is an advantage to this sort of planning. You can save time. This doesn't mean an experienced draftsman follows this routine, but then he doesn't need to. He does his layouts mentally and so will you, in time.

When using the pencil, make sure the pencil point is always sharp. With use, the point tends to become wedge-shaped, and this will result in lines that gradually become thicker. The same is true of the lead you use in your compass. One of the factors a professional and experienced draftsman uses in evaluating a drawing is uniformity of weight of line.

You can shape the lead in your compass by using the sandpaper pad. Twirl the pencil to rotate the lead, but do not shape it by flattening one side (Fig. 3-22). Be sure to bring the lead to a fine point. Avoid a wedge shape since this will result in a broad line (Fig. 3-23). Figure 1-24 in Chapter 1 shows the wrong and right shapes for the compass lead. Use the point illustrated in Fig. 1-24.

If the work you are about to do requires the use of both a pencil and a compass, you can insure uniformity of lines by using the same lead for both. The way to do this is to sharpen a pencil to a sharp point. Cut off the lead and use it to replace the lead that normally comes with the compass. You will need to sharpen the pencil again to obtain a point, but now the pencil and the compass will be using identical leads, with both leads shaped to identical points.

If the drawing you are producing is a small one that does not require much pencil work, you may be able to do the entire drawing

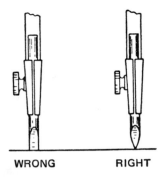

Fig. 3-23. Wrong and right shapes for a compass lead.

without additional pencil sharpening. However, develop the habit of watching both the pencil point and the thickness of the line that the pencil produces. It is easier to sharpen a pencil than to erase lines and redraw them. It also results in a cleaner and more professional looking drawing.

INK WORK

Ink has many advantages. It is much more permament than pencil. Certain types of drawing surfaces accept ink only, and if the drawing is to be reproduced by some photographic process, the prints may very well look better and sharper when ink is used on the original.

It is a little more difficult to work with ink than with pencil. The reason you may originally feel more confident in working with pencil is that you are more accustomed to it. Working with a ruling pen is a new experience.

Filling A Pen

When trying a new pen for the first time, use a clean, dry cloth and rub all its metal surface, inside and out, to remove any traces of oil that may have adhered to the metal surface. On the side of the pen you will find a small, knurled screw. This screw controls the size of the pen opening. If you make the opening too wide, the pen will not hold the ink and the ink will drop to your working surface.

Don't try to put ink into a pen that isn't clean. The interior surface of the pen should show a metallic surface. If it is covered with dried ink, it will look dull black. Special ink remover is available for cleaning dried ink, but if the ink has been allowed to accumulate over a period of time you may need to let the pen soak in the remover to dissolve the ink. You can use a mixture of household ammonia and water.

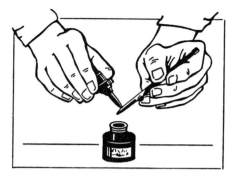

Fig. 3-24. How to load a pen with ink.

The best procedure is always to clean a pen after you are finished using it. At that time the ink is easily removable with a soft cloth.

The average bottle of ink for drawings contains 2 ounces and the stopper may come equipped with some sort of device to let you transfer the ink from the bottle to the pen. Figure 3-24 shows the technique for filling the pen. Never fill a drawing pen by dipping it into an ink bottle.

Hold the pen over the open bottle so that if any ink spills, it will go back into the bottle where it belongs. Insert the ink filler between the nibs of the pen so that you get about 1/4 inch ink level. Don't try to overload the pen since a small abrupt movement could discharge all or part of the ink reservoir, causing ink spillage or a blot. If the stopper of your ink bottle does not come equipped with any means for transferring ink to a pen, use a toothpick, or an ordinary, old-fashioned pen point. These are available in art and drafting supply stores. You can also use a small eye dropper for this purpose.

If a bottle of ink can possibly fall over, it will. While the usual 2 ounce bottle of drafting ink is quite small, it can be upset. You can prevent this by using a weighted base. Some come equipped with a lever-operated filling attachment to make filling the pen easier and quicker.

India ink evaporates rapidly. Keep the bottle of ink corked when you aren't using it. Do not keep the stopper out any longer than necessary. If the ink does thicken, however, you can thin it by adding a mixture of household ammonia and water. Use 4 parts of ammonia to 1 part water. Distilled water is preferable. The usual and most used color for drafting ink is black, but you can also get it in different colors.

Using The Pen

When you use a pen to draw a line, hold the pen so that the adjusting setscrew is pointed away from you. Figure 3-25 shows two drawings, illustrating how to hold the pen. As you can see in

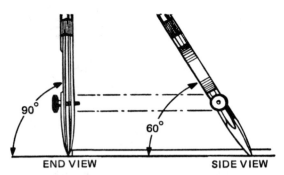

Fig. 3-25. When ruling a line, incline the pen in the direction in which you are drawing.

the drawing at the right, tilt the pen slightly in the direction you are drawing. Thus, in this drawing at the right we are moving horizontally across the paper, from right to left.

As shown in the drawing at the left, the pen is perpendicular to the bottom edge of the sheet. To get used to the idea of producing inked drawings, practice drawing inked lines. The best place to practice is on scrap, and not on an actual drawing. Try drawing lines of various thicknesses.

Professional draftsmen always draw sample ink lines on scrap paper before starting work on drawings. Because they are experienced, they can usually look at a line and evaluate its thickness. However, until you acquire this ability, there is a little drafting trick you can use as a drawing aid. After you have adjusted the screwhead of your ruling pen, put a mark on it. After you load the pen with ink, always rotate the screwhead until the mark is in its original position.

You can also mark the pen for different amounts of line thickness, as shown in Fig. 3-26. As an example, when the mark is at "b," the

Fig. 3-26. Index marks on a ruling pen will help you locate the position of the screwhead for lines of different thickness.

Fig. 3-27. Always draw an inked line directly over the pencil line.

pen may be adjusted for a fine line; when the mark is at "a" a medium line results; and when the mark is at "c," you will get a thick line. Do not change line thickness while a drawing is in progress, unless there is some special reason for doing so.

When drawing inked lines, however, give the line a chance to dry. If the line isn't dry, it may be smeared by movement of the T-square. A line that is not quite dry will run if another inked line is drawn through it.

When inking pencil lines (see Fig. 3-27) make sure that the center of the ink line is drawn directly over the center of the pencil line. And, as in the case of pencil lines, make sure that the weight of line is uniform.

Inking Problems

Both pencil and ink have problems of their own. Although India ink can be erased, it requires much more effort than erasing pencil

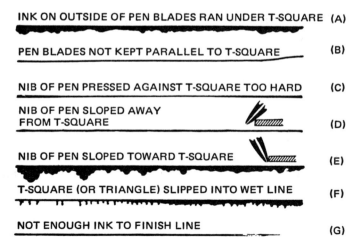

Fig. 3-28. Various inking faults. (A). Ink on outside of pen blades ran under T-square. (B). Pen blades not kept parallel to T-square. (C). Nib of pen pressed against T-square too hard. (D). Nib of pen sloped away from T-square. (E). Nib of pen sloped toward T-square. (F). T-square (or triangle) slipped into wet line. (G). Not enough ink to finish line.

lines. Extensive erasing of inked lines can produce a hole in the paper or can spoil the paper, making it too rough for further inked lines.

Figure 3-28A shows a common inking fault. Here the draftsman tried to rule a line using his T-square, and ink ran under the T-square. This not only ruins the drawing completely, for it is almost impossible to erase such a large ink smear. It is equally difficult to remove the ink from the T-square surface. This problem arose because of the presence of some ink on the outside metal surface of the ruling pen. If, for some reason, you do see some ink on the outside of the pen, remove it very carefully with a dry cloth.

Figure 3-28B shows another common trouble. If you will examine the right side of the line, you will see it moves upward. This fault occurs when you try to draw a very long line with the T-square. When you draw a line you must keep the blade of the pen parallel to the edge of the T-square. If the pen wiggles, that is what you will get — a line with a wiggle.

To avoid the situation in Fig. 3-28B, a novice draftsman will try to keep the points of the pen too tightly against the edge of the T-square or triangle. The object here is to keep the pen from moving back and forth. Again, the rule is that the pen must be vertical to the drawing paper, and not even slightly tilted to take advantage of possible support from the T-square or triangle. The result, as shown in Fig. 3-28C, is a line that has varying thickness.

You will get the same results — a non-uniform line — if you let the nib of the pen slope away from the edge of the T-square or triangle. Figure 3-28D shows the result.

If you force the nib of the pen against the edge of the T-square or triangle, the pressure may force ink out of the sides of the pen. Ink will then flow from the pen beneath the T-square or triangle, producing the smear shown in Fig. 3-28E.

India ink does dry fast, but you must give it enough time to dry. If you move your T-square or triangle into the "not quite dry as yet" line, the result will be an ink smear (Fig. 3-28F).

When you ink a drawing, always take a look at the amount of ink remaining in the pen before drawing a line. Otherwise you may find you do not have enough ink left to finish the line (Fig. 3-28G). While you can rule over the line once again, unless you have quite a bit of experience, you will either not be able to superimpose the new line on the old, incomplete one, or else you will thicken the line.

Practice Exercise

- Examine your ruling pen and, turning the screw on the side, note its effect on the points of the pen.

- Using India ink, fill your ruling pen so the ink forms a reservoir of about 1/4 inch. Mount a sheet on your drawing board, and with the help of a T-square, try drawing a few horizontal lines.
- Adjust the screw and draw a number of lines, each having a different thickness.
- With the help of a triangle and T-square, draw a number of vertical lines. After the lines have dried, try drawing some horizontal lines so they intersect the vertical lines.
- Using a triangle, draw several slant lines in ink.

Tips On Handling A Pen

After practicing for a while with a pen, you will find yourself able to draw lines without spilling ink. Here are a few rules that will increase your ability to handle a drawing pen.

- Always be sure to wipe all traces of ink from the outside of the blades. After you fill a pen with ink, some of the ink may be on the outside of the blades. You can remove this ink by carefully using a soft cloth. Be careful not to touch the cloth to the inside of the pen. If you do not remove the ink from the outside of the blades, you will smear the T-square and also your drawing.
- If you finish a drawing, do not put the pen down without wiping away all of the ink remaining in the pen. Ink that is allowed to stay in a pen will evaporate, but will leave behind a hard black ink surface that will be difficult to remove.
- Never fill a pen that already has ink in it with new ink. Always wipe the pen clean first.
- Clean the pen thoroughly once a day. Make sure no ink marks remain between the inside surfaces of the blades. Do not try to scratch or file away any ink residue that remains.
- When doing an inked drawing, you will find your work going easier and faster if you draw all horizontal lines first. Then do the vertical lines. Using your T-square, start at the top of the work page, and then draw all the necessary horizontal lines until you reach the bottom of the drawing. Wait a few moments for the ink of the last line to dry. Using a triangle resting against the T-square, draw all the vertical lines.

  Work from the right side of the work sheet to the left side. If you work from left to right instead, the movement of the T-square over the work may smear a line that isn't quite dry. Never draw intersecting lines in ink until the ink is thoroughly dry.
- If you must erase an inked line, wait until the line is dry.

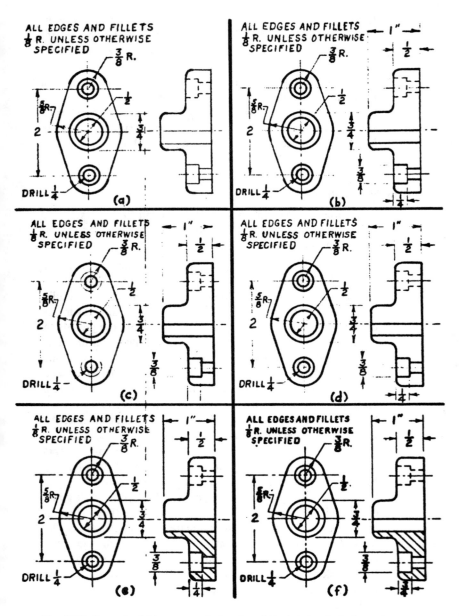

Fig. 3-29. Steps in inking a pencil drawing. (A). Ink in all circles and arcs.
(B). Adjust the pen accordingly when inking lines. (C). After the horizontal
lines have been inked, ink in the vertical lines. (D). Wipe the T-square from
time to time. (E). Ink in all slant lines using the T-square and a triangle.
(F). Ink in all lettering.

Inking a Drawing

To ink a drawing having just horizontal and vertical lines, follow the procedure outlined earlier. It is unusual, though, for a drawing to have just horizontal and vertical lines. A drawing will also have arcs, rounds, fillets, curves and slant lines. Here is a procedure you can use that will help you in two ways. It enables you to complete your work as quickly as possible and helps you avoid the possibility of smearing inked lines.

— The first step in inking should be to ink in all circles and arcs. See Fig. 3-29A. You may find yourself having a tendency toward making the inked line slightly larger than the pencil line. Always ink right over the center of pencil lines.

— Ink all horizontal lines from top to bottom. In some drawings you may deliberately want some lines to be thicker than others. If the horizontal lines are not all of the same weight, set your pen for one line thickness and then ink in all lines having this thickness. Then readjust the screwhead of the ruling pen for the line of the next thickness, and draw all of these. Some draftsmen prefer to ink all lines in order, one after the other, and adjust the pen accordingly. This takes quite a bit of experience and is done by draftsmen who are so familiar with the ruling pen they can adjust the screwhead of the pen almost automatically. See Fig. 3-29B.

— After the horizontal lines have been inked, ink in the vertical lines, working from right to left. While working you may find it good practice to wipe the T-square from time to time just to make sure it hasn't picked up any lint or dirt. Use a clean, dry cloth. Do not use the cloth you use for cleaning your ruling pen. (Fig. 3-29C and Fig. 3-29D).

— Ink in all slant lines using the T-square and a triangle (Fig. 3-29E). Before you work with the triangle, wipe it with a clean, dry cloth. When drawing slant lines try to space them as evenly as possible. Since you will be following your pencil lines, you will get uniform spacing if the pencil lines have been properly spaced.

— Ink in any lines that have not been covered in the steps listed above. Examine the drawing carefully to make sure that no lines have been omitted.

— Your final step will be to ink in all lettering (Fig. 3-29F). After you have completed the drawing, take a good look at it. No pencil lines should be visible, but if you can see any, erase them. Remove any unnecessary lines. Use a soft brush to remove any material left by erasing.

- Be sure to read all of the lettering, including any notes you may have made. While an error in spelling does not alter the accuracy of the drawing, it doesn't inspire confidence. It may even raise the thought that if you could be careless in this respect, then possibly your drawing may have some errors not so immediately obvious.
- Every drafting department has its own procedures. In some cases the drawing must be submitted to a checker. In others, the rule is to make duplicate copies by a process such as Xerox. In others a time record must be filled in to indicate the amount of time required for the drawing.

ERASERS

There are many types of erasers you can use in your work. Draftsmen generally have personal preferences and use different kinds of erasers for various kinds of work.

Figure 3-30 shows just a few of the many erasers available. In the upper left-hand corner is a plastic eraser for cleaning plastic surface drawing mediums. The eraser shown at the top right is red and is used for removing pencil lines. It is a general purpose eraser and can be used on cloth or paper.

At the lower left in Fig. 3-30 is an eraser with which you are probably already familiar. It can be used for erasing pencil and ink on linen and paper. This particular eraser is available in various grades. The softest generally has a pink color. Intermediate erasers are usually red, while the hardest of these erasers is white.

Finally, the eraser shown at the bottom right is known as an artgum eraser or, more simply, as a gum eraser. It is used for cleaning smudges or finger marks from the surface of a drawing.

Erasers can change with age, some more, others less. Some tend to become hard and brittle. When they are in this condition, even though they may have been barely used, they smear more than they erase. They can produce smudges that are difficult to erase and so it

Fig. 3-30. Some of the many erasers available for drafting.

is an act of economy to get rid of them. Sometimes, "art" stores will offer erasers at a bargain disposal price. Since these may be very old stock, they may not be quite the value they seem to be.

Using Erasers

The purpose of an eraser is to remove a pencil or ink line without damage to the work surface. Erasures are successful only if you can remove the line or lines which are in error, and you are able to replace them with lines which are correct. The harder the eraser, the more you are likely to do some damage to the work surface. This does not mean you should always use the softest erasers available, for these may be too soft to remove the unwanted lines.

Using erasers requires judgment, patience and skill. If erasing roughens the work surface, you may find it impossible to use the surface for drawing. Where the drawing is fairly elaborate and has required a lot of time, it may be possible to put in a patch. Whether or not this will be permitted depends on the rules of your company or on your own attitude toward your work.

Some draftsmen use a single edge razor blade to help in removing stubborn ink lines. Again, this requires some know-how. If you would like to add this technique to your growing skills, try it on inked lines drawn on scrap paper or cloth.

The Dusting Brush

After you have completed erasing, you will find you have eraser crumbs and possibly bits of paper over the drawing surface. A fast and easy way to get rid of them is to use a dusting brush, similar to the one shown in Fig. 3-31. Do not remove eraser crumbs by wiping the drawing surface with your palm. The crumbs will not be clean and will contain embedded particles of ink and graphite. These will smear the drawing, requiring still further erasing. Do not blow across the face of the drawing. Aside from being an unsanitary practice which may repel those who must handle your drawings, it can also cause smearing.

The Electric Erasing Machine

Electric erasing machines are often used in drafting rooms where a large number of drawings must be produced in the shortest possible

Fig. 3-31. A dusting brush is a fast way to clean drawing. A soft bristle brush is preferable to the cheaper type using plastic.

time. The erasing machine is a fast way of erasing, but it requires some skill in its use for it is very easy to rub a hole in either paper or cloth. The electric eraser must be handled very gently, as it can bite into the surface of paper or cloth.

If you can get an opportunity to practice with an electric eraser, do so, not on a finished drawing, of course, but on scrap. Erasing correctly depends on the kind of paper or cloth used, whether the lines are penciled or inked, the type of eraser used in the electric machine (the erasing tips are interchangeable), and the skill of the draftsman.

The Erasing Shield

One of the problems in erasing is to remove unwanted lines without affecting other lines. If the lines of a drawing are widely spaced, there is no problem. However, you will sometimes find lines to be erased positioned immediately adjacent to other lines.

To overcome this difficulty, you can use a very inexpensive tool known as an erasing shield. The erasing shield is a rectangular plate of thin steel or plastic with rectangular areas and holes of various sizes cut through it. To use the shield, select an opening that exposes an unwanted line, but with the remainder of the shield covering and protecting the other lines. You will find the punched edges of the shield will cut away the smudged edges of the erasers you use, resulting in a clean rubbing edge.

An erasing shield, like any other object you may put down on the work surface of your drawing, can pick up dirt, particularly erasing crumbs, and is then capable of smearing the work. After you finish using the erasing shield, wipe it clean with a dry cloth.

Before you use an eraser, examine its surface. If the surface looks dirty or dull, you can clean and freshen it by rubbing it across some fine sandpaper. Then, rub the eraser across a piece of clean scrap to remove any sandpaper particles. With a clean shield and an eraser surface in good condition, you have at least met some of the conditions for erasing successfully.

Other Erasing Techniques

Various erasing methods have been evolved since drawing errors seem to be inevitable. An error may be an actual mistake in drawing a line, drawing a line poorly, or drawing a line where no line should exist. Errors in drawings may also be due to changes in instructions.

In addition to the erasers and the erasing machine, various cleaning compounds are available. These include pulverized gum eraser particles which can be squeezed onto the drawing from a plastic squeeze bottle. You can also obtain gum eraser granules packaged in cloth bags made of soft mesh. When you rub the bag over the drawing, the granules will sift through the bag.

As mentioned earlier, you can remove ink lines with a razor blade. A single edge type is preferable since the blade backing makes the blade stiff and more easily controlled. Double edge blades are too flexible. Use a brand-new blade since a sharp edge will make the ink more easily removable.

Some draftsmen prefer a steel blade made especially for the purpose of erasing. The long handle makes it easier to control the blade, and it also results in less finger tension and fatigue.

Erasing involves scraping or rubbing or some combination of both. Erasing is more generally used for pencil lines; scraping is for ink lines.

MAKING CHANGES OR CORRECTIONS

Sometimes, after completing a drawing, you may be asked to make changes or corrections. If the changes are extensive, it may be easier and quicker to do a completely new drawing. A change may also mean some rearrangement of parts of the drawing. You, or the drafting supervisor may feel it advisable to start all over. Sometimes the change will be a simple one, but there will be no way in which to accommodate the change on the existing drawing. However, if the correction is simple and can be made to fit with the existing drawing, you may be able to save drafting time with a revision.

To make a change, put a sheet of tracing paper over your drawing. Then make the required changes right on the tracing paper, using a pencil, a compass and a T-square, just as though you were doing an original drawing. Do not make just the changes alone on the tracing paper. Extend your lines a bit so there is some overlap between the correction lines and the original drawing lines.

After you have done this, remove the tracing paper and erase the area on the original drawing that is to be corrected. Now put the tracing beneath the original drawing and put both on a light box. With the light turned on you will be able to see the lines of both the tracing and the original drawing. You can now trace the corrections onto the work drawing. In doing this, make sure the newly traced lines and the original lines merge. At this time any variation in line thickness will show up and reveal the patch for what it is. A correction on a drawing is a good correction only if it cannot be recognized as such.

Using a light box is helpful, but may not always be desirable or possible. In that case, make a rough sketch of the proposed change, but draw it in the same scale as the original drawing. With the help of such a sketch you will be able to decide if it is possible to make a change directly on the original, and whether it will fit into place.

TRACING ON CLOTH

Drawings are made on cloth for several reasons. A drawing on cloth is much more permanent than one made on paper. The drawing may need to be kept for a long time. Drawings on cloth can also tolerate much more handling and abuse than paper drawings. For example, the drawing may need to withstand considerable handling during some manufacturing process. The drawing may be passed along from hand to hand, repeatedly rolled and unrolled. Despite all the work and effort you put in on it, the drawing may spend its days and nights tossed into some dirty cubbyhole. Finally, cloth drawings are used where precision is essential.

Preliminary Steps

If you have been asked to make a drawing on cloth, first make sure the pencil drawing from which you will trace the ink drawing is accurate in every detail. If, after you have completed your work on the cloth you find an error in the original drawing, you will see how difficult it is to erase.

Tracing cloth is not as transparent as paper. To save yourself some eyestrain and to avoid overlooking lines on the original drawing, you may find it helpful to go over the pencil lines with an H pencil.

One of your difficulties may be to locate the center point of circles and arcs. Use the needle point of your compass to make the center points a bit larger. Since this may damage the original work drawing, some draftsmen prefer to emphasize the center point by drawing small dark circles around all centers. This makes it possible to identify center points easily.

Preparing The Work Surface

Cloth used for drawing has two sides that are different. One side is shiny; the other is dull. The dull side is the working side. The dull surface of the cloth has a film that repels ink. This film can be removed by dusting a powder, known as *pounce,* on the cloth and rubbing it into the surface of the cloth with a clean rag. When you have finished doing this, brush off any excess pounce.

Do not use water on tracing cloth. Always make sure the cloth is kept in a clean, dry place. The working surface of the cloth contains starch. Water will remove the starch and destroy the work. Finally, to get accustomed to working on cloth, be sure to practice first on some scrap pieces.

Order Of Inking

To draw ink lines on cloth, follow the same procedures given previously for drawing on paper. There is a precaution, though. Tracing cloth can change its shape, depending on weather conditions. For this reason you should ink only a section of the drawing which you can finish at one sitting. This means, then, that you may need to alter somewhat the preferred technique of drawing all horizontal lines and then all vertical lines. Select an area of the drawing which you know you can finish in a certain time. Treat this section as though it were a complete drawing. Within this limited area, you can still follow all the inking rules, doing the horizontal lines first and then the vertical lines and curves.

If you start to ink circles, you must do all of the circles on the entire sheet. Do not start working on circles if you find that you can do only part of them.

Erasing On Tracing Cloth

Use an ink eraser to remove errors in ink lines on tracing cloth. It takes some skill to erase and not expose the threads of which the cloth is made. If you remove the cloth surface and then try to ink over it, you will find the ink spreading over the cloth in the form of a blotch. To keep this from happening, rub soapstone over the erased area. The soapstone can be put on directly, or you can use powdered soapstone and rub it on with a knife. Rub it in with your finger and then polish it with the back of a fingernail.

THINGS TO DO

- Fill your ruling pen so that the reservoir of ink in the pen extends up about 1/4 inch from the tip of the pen. Experiment with the adjustment screw on the side of the pen and note that it can make the two tips (the nibs) of the pen move together or apart. Note at the same time that this action forces the ink up or down in the pen.
- Fasten a sheet of drawing paper to the board, using the techniques previously described. Start near the top of the paper and draw a succession of horizontal lines, moving the T-square down as you do so.
- After you have drawn these lines, draw a series of lines which are thinner and thicker.
- Now, with the help of a triangle, draw a series of vertical lines starting at the right side of your work paper and gradually moving over to the left side after you complete each individual line.

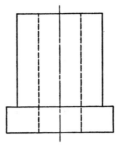

Fig. 3-32. Three types of lines are used in this drawing.

- Using the slant edge of your triangle, draw a series of lines, trying to space them equally from each other.
- If you have an ink compass, fill it and practice drawing circles. Then draw a series of arcs of various sizes.
- Try erasing some pencil lines and then some ink lines. After you have erased, try drawing new lines in pencil and in ink over the erasures. The more you practice, the more skilled you will become. After you are finished, wipe all ink out of the pens. Inside and outside surfaces of the pens should look new and shiny after you are finished.
- Without using a ruler, transfer the measurements of the drawing shown in Fig. 3-32 to a work sheet and do the drawing. There are three types of lines used in this drawing. Identify them all.
- Without using a ruler, transfer the measurements of the drawing shown in Fig. 3-33 to a work sheet and produce a duplicate. Do the drawing in pencil. Identify all lines.
- Without using a ruler, transfer the measurements of the drawing in Fig. 3-34 to a work sheet and draw a duplicate. Do the drawing in pencil. Identify all lines.
- Ink in the drawing you did by following Fig. 3-32.

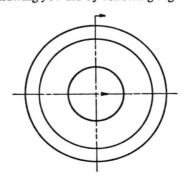

Fig. 3-33. Practice drawing these circles and lines.

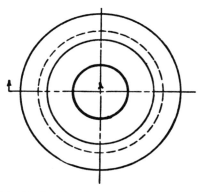

Fig. 3-34. Ink in this drawing when you have finished duplicating it.

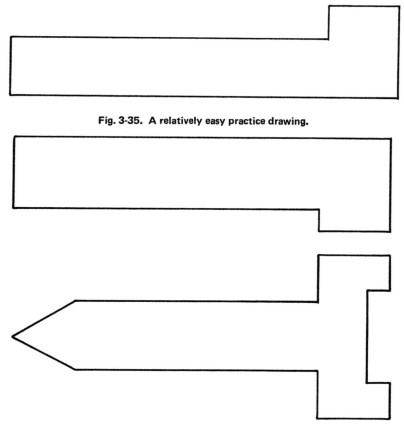

Fig. 3-35. A relatively easy practice drawing.

Fig. 3-36. Note the positioning of these two drawings
with respect to each other.

- Without using a ruler, transfer the measurements of the drawing in Fig. 3-35 to a work sheet and draw a duplicate. Do the drawing in pencil and then ink it.
- There are two drawings in Fig. 3-36. Note how they are positioned with respect to each other. Transfer both drawings to a work sheet and produce a duplicate, in pencil. Ink the drawings when you are finished.
- Practice drawing arrowheads.

SUMMARY

- A center line consists of long and very short dashes. The dashes are evenly spaced and each center line starts and ends with a long dash. Center lines are used to indicate the center of a symmetrical object.
- Visible lines, also known as outlines, are solid and thicker than center lines. Visible lines define the outline of the object being drawn.
- Concentric circles consist of one circle (or more) drawn inside another circle.
- Hidden lines represent a part of the object that cannot be seen. They are indicated by short dashes.
- The thickness of a line is known as its weight. Lines used in drafting have three different weights — thin, medium, and thick. Thin lines include center lines, dimension lines, leaders, long break lines, extension lines and sectioning lines.
- Two types of lines are needed for the dimensions on a drawing. These are extension lines and dimension lines. Dimension lines end in arrowheads.
- A leader is a line used to call attention to some property of the object. The leader, ending in an arrowhead, touches the visible line.
- Break lines are used to indicate a continuous length of identical material. The purpose of a break line is to save drawing space.
- Phantom lines are used to indicate one or more possible positions that may be taken by a moving object.
- Sectioning lines are used to indicate the exposed surface of an object.
- A cutting plane line shows the position of a surface being exposed.

Chapter 4
Lettering

You will need to use lettering on just about all your drawings. Lettering consists of capital letters, small letters and numbers. Capital letters are generally called upper case, while small letters can be identified as lower case.

LETTERING AIDS

Most draftsmen make use of various lettering aids. These are lettering guides and their purpose is to help the draftsmen produce professional lettering in minimum time. There are various types of lettering guides, such as Leroy and Wrico. The Leroy, shown in Fig. 4-1A requires the use of a special lettering pen or scriber and a template. The Wrico also uses a template.

Either the Leroy or the Wrico are frequently used by professional draftsmen. At the start, however, you can try a Braddock-Rowe triangle or an Ames lettering guide. These are simple, relatively inexpensive devices to help you draw guidelines for the letters. Both the Braddock-Rowe and the Ames are used for drawing horizontal guide lines only (Fig. 4-1B). They are equipped with a number of through holes for the insertion of a pencil point. When in use they are supported by the T-square while the guide and inserted pencil are moved horizontally across the work sheet. Their big advantage is that they supply guide lines that are correctly spaced away from each other. You can also do lettering without the help of any of these aids.

One of the advantages of the Leroy or Wrico systems is that they permit you to letter directly in ink. However, if you plan to do freehand lettering it is better to work in pencil. Use either an F, H or 2H. Make sure it is correctly sharpened, but this does not mean the pencil should have a needle point. You can dull the point, if it is too sharp, by twirling it on scrap paper.

You will find it helpful to begin working with lettering using cross-hatch graph paper. The vertical and horizontal lines of this paper will act as guides for the various parts of the letters.

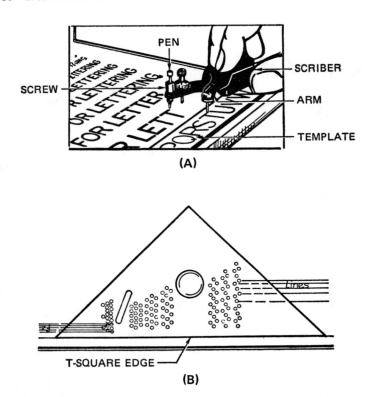

Fig. 4-1. (A). The Leroy lettering guide (B). The Braddock-Rowe triangle.

Lettering may be either vertical or slanted. All lettering, whether vertical or slanted, used in drafting, is called Gothic. Slanted letters are also known as inclined or italic.

VERTICAL NUMERALS

Figure 4-2 illustrates how to draw numbers. If you will examine the numbers, you will see each has one or more arrows and that these arrows are numbered. The arrows indicate the direction of movement of your pencil while the adjacent numbers specify the order in which the lines may be drawn. Note also that the numbers from 0 to 9 are not arranged in numerical sequence.

Numbers shown in the second row are made of curved lines. Those in the upper row can all be made with straight lines. Digits 4 and 7 can be drawn with one of the lines having a slight curve. While this lettering technique is used, it is better to work with straight lines only. Copy these numbers on graph paper and try to have those you

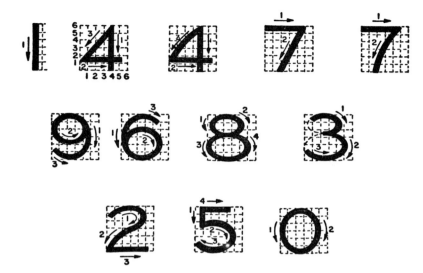

Fig. 4-2. Technique for drawing vertical numerals. Start practicing with the upper row first since curved numbers are more difficult to draw than those having straight lines only. The stroke sequences shown here is not a rule, but a suggestion. Change it if you wish.

draw resemble those in the illustration as closely as possible. It takes quite a bit of practice to become competent in lettering, so don't expect to be able to acquire this skill overnight. While you can start by drawing these numbers rather large, keep making them smaller until they are about 1/8 inch high, spaced 1/8 inch between rows. The total height of fractions is twice that of integers (whole numbers).

The numbers shown in Fig. 4-2 look quite thick. This is due to the fact that they have been "blown up" to a number of times larger than their original and correct drawing size. You may find it helpful, as a start, to put a tissue overlay on top of Fig. 4-2 and trace the numbers. It won't be necessary to make the numbers as thick as those shown. Instead, trace the outlines of the numbers. While you do so, try to memorize the steps you take in doing these drawings. When you draw line 1 of a number, make yourself conscious of the fact that you are drawing line 1 by repeating that number to yourself. The stroke sequence, however, is a suggestion, not a rule. Do the same with line 2 and following lines. After practicing with the overlay sheet, use a sheet of graph paper to see how closely you can come to reproducing the numbers in Fig. 4-2.

When you draw numbers or letters make each line a single stroke; that is, do not try to "sketch" the numbers by using a series of very

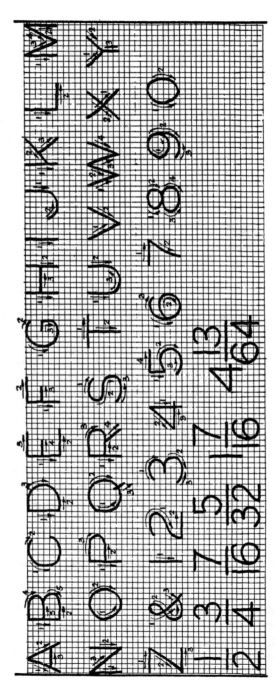

Fig. 4-3. Stroke sequence for upper case (capital) letters and fractions. Sequence of strokes is for right-handers. Those who are left-handed should work out the sequence that is most comfortable for them.

short lines. If you make a mistake when drawing a line, stop and erase what you have done and then start the line all over again. Do not try to erase a small section of a faulty line and then fill it in. Each number and each letter must give the impression of being a single, continuous line of uniform thickness.

You will be able to complete some numbers with a single pencil stroke. The number 1, for example is, is a single stroke number. The number 7 is also a single stroke number with the digits shown on the number 7 in Fig. 4-2 indicating the sequence of pencil movement. In the number 5 you will need to lift your pencil from the paper after completing the first, second and third movement of the pencil. You will then need to do step 4, the short horizontal line at the top of the number.

UPPER CASE LETTERS

After you have practiced drawing vertical numerals, move ahead to drawing upper case letters as shown in Fig. 4-3. The lettering in Fig. 4-3 may seem like a lot to learn and practice but it isn't as much as it seems. Consider the letter C, for example. It has the same shape as the letters O and Q. Many of the letters have straight lines — letters such as E, F, H, I, L, N and T. If you can draw a straight line for one letter, you can draw a straight line for any other letter. Using graph paper, you will find straight lines, whether horizontal or vertical, much easier to draw than curved lines. For this reason you might want to practice straight line letters first. Once you have the confidence supplied by drawing such letters (or numbers), move on to those that have curves. Do not try to use a French curve as a lettering aid.

While graph paper is suitable for first attempts at lettering, you will ultimately need to abandon it since drafting plates are made on unruled sheets. Your goal should be to draw letters and numbers with the minimum number of guide lines.

Figure 4-4 shows vertical lettering using horizontal guide lines only. Copy this illustration. Begin by drawing horizontal guide lines with the help of your T-square. Note the way in which the letters are spaced. You will find that each letter is like an individual and requires an amount of space depending on its shape. The letter I, for example, doesn't take up as much room as the letter H. Draw the letters 5/16-inch high. However, if you have any difficulty in producing letters in such a small size, do what we have done in Fig. 4-4. The upper part of this illustration shows letters that are fairly large. After you have acquired experience with letters of this size, try drawing them smaller as shown in the lower half of Fig. 4-4.

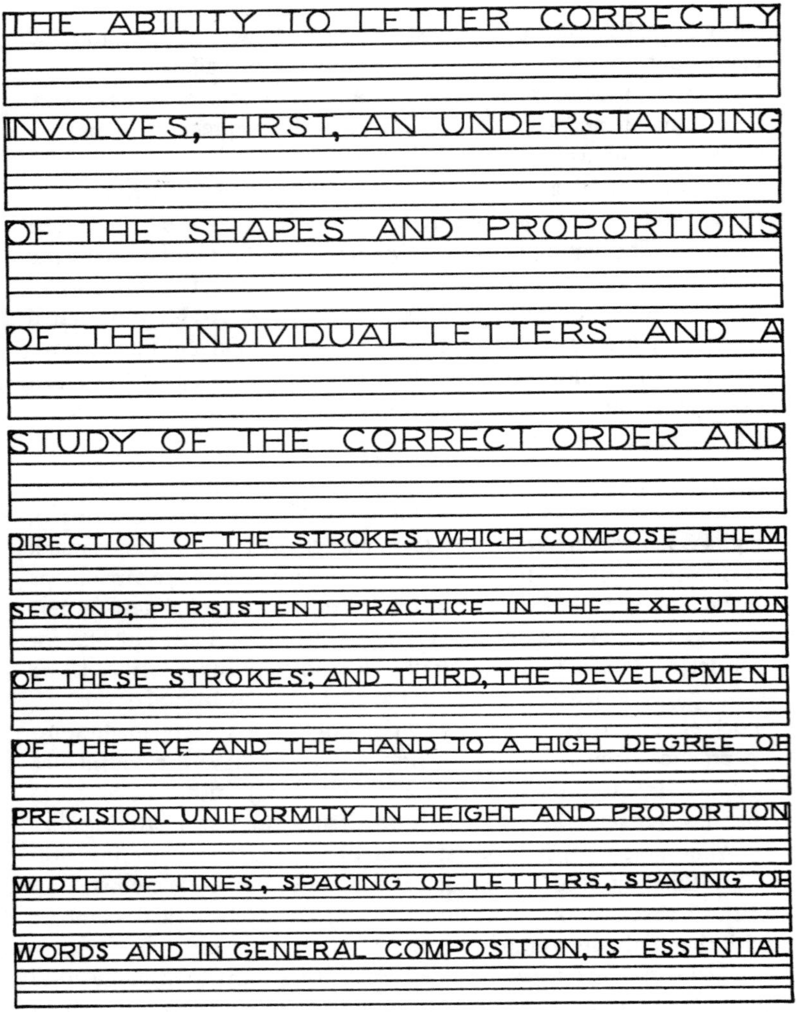

THE ABILITY TO LETTER CORRECTLY

INVOLVES, FIRST, AN UNDERSTANDING

OF THE SHAPES AND PROPORTIONS

OF THE INDIVIDUAL LETTERS AND A

STUDY OF THE CORRECT ORDER AND

DIRECTION OF THE STROKES WHICH COMPOSE THEM

SECOND; PERSISTENT PRACTICE IN THE EXECUTION

OF THESE STROKES; AND THIRD, THE DEVELOPMENT

OF THE EYE AND THE HAND TO A HIGH DEGREE OF

PRECISION. UNIFORMITY IN HEIGHT AND PROPORTION

WIDTH OF LINES, SPACING OF LETTERS, SPACING OF

WORDS AND IN GENERAL COMPOSITION, IS ESSENTIAL

Fig. 4-4. Vertical lettering using horizontal guide lines only.

You will note that in Figures 4-3 and 4-4 that we have used capital letters only. The reason for this is that capital letters are easier to draw than lower case letters.

INCLINED LETTERS

You may find it a bit more difficult to draw inclined letters. The problem here is to make the angle of slant the same for all letters.

Fig. 4-5. Formation of inclined upper case and lower case slant letters.

The angle of slant depends on your personal preference. Some drafts-men use an angle of 67 1/2° with the horizontal. Others use an angle of 75°. Again, it isn't the amount of slant that is so important as having uniformity of slant.

The inclined Gothic lettering in Fig. 4-5 shows how the letters are formed. Unfortunately, you will not have slant graph paper available to help you, but you may find it useful to draw a succession of slant lines at random after you have completed your horizontal guide lines.

You will also find it easier to use 75° as your slant angle since, as indicated in Fig. 4-6, you can draw slant guide lines with the help of a pair of triangles — a 30° and a 45° triangle. Draw the guide lines very lightly. After you have finished your practice plate, erase them carefully. In Figure 4-7 you have the complete alphabet of slant letters, in upper and lower case, together with whole numbers and fractions. The fact that the upper row consists of thick letters while the other rows are much thinner is of no significance. The thickness is exaggerated and just means that some pencils, depending on type, will produce thicker lines than others.

Whether you draw vertical or slant letters and numbers may be a personal matter or the decision of the company employing you as a draftsman. However, when drafting as a hobby or for your own use, you will probably want to use the vertical style. While Figure 4-8 shows both vertical and slant letters, you will rarely find both types used on the same drawing. If you are planning on a series of plates, all relating to the same object, it would be better to have all the lettering uniform, both as to size and style. This uniformity extends to upper and lower case. Do not intermix them on one drawing, and then use just upper or lower on the others. Consistency is a good drafting habit.

LOWER CASE LETTERS

Lower case or small letters, oddly enough, may be more difficult to draw than upper case. For lower case letters you will need to draw four horizontal guide lines as shown in Fig. 4-9. The body of each letter is contained within two of these lines. The upper is called

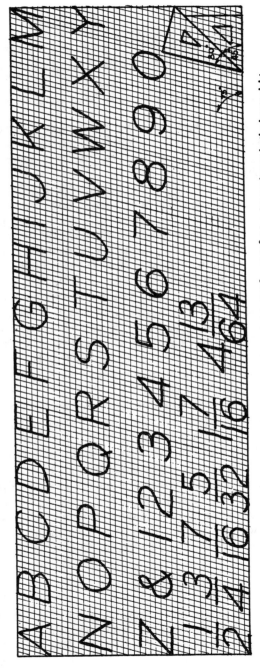

Fig. 4-6. You can set up a series of slant guide lines with the help of 30° and 45° triangles as shown in the lower right.

ABCDEFGHIJKLMNOPQRSTUVWXYZ

1 2 3 4 5 6 7 8 9 0 $\frac{1}{2}$ $\frac{3}{4}$ $\frac{5}{8}$ $\frac{7}{16}$

ABCDEFGHIJKLMNOPQRSTUVWXYZ

1 2 3 4 5 6 7 8 9 0 $\frac{1}{2}$ $\frac{3}{4}$ $\frac{5}{8}$ $\frac{7}{16}$

ABCDEFGHIJKLMNOPQRSTUVWXYZ

1 2 3 4 5 6 7 8 9 0 $\frac{1}{2}$ $\frac{3}{4}$ $\frac{5}{8}$ $\frac{7}{16}$

ABCDEFGHIJKLMNOPQRSTUVWXYZ   abcdefghijklmnopqrstuvwxyz

$.498 + .000$ .498
$.498 - .002$ .496

$.498 - .002$

Fig. 4-7 Slant letters, whole numbers and fractions.

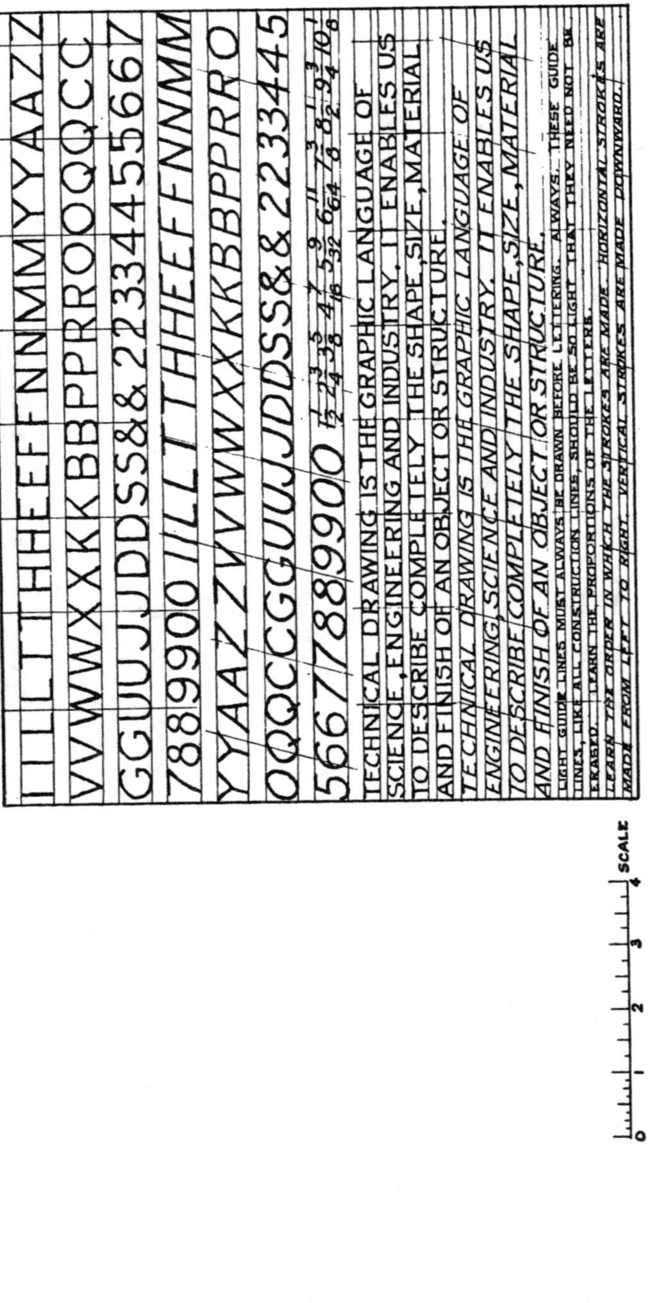

Fig. 4-8. Vertical and slant letters and numbers in various sizes. While horizontal guide lines are used for all, note also the random use of vertical guide lines for both vertical and slant lettering.

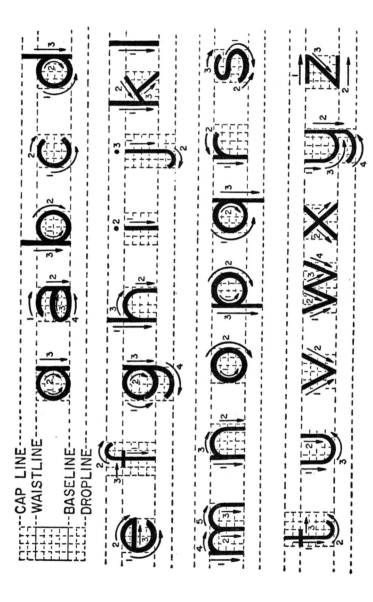

Fig. 4-9. Method of drawing lower case letters.

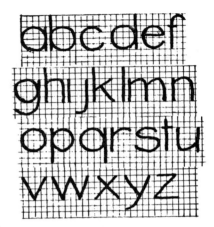

Fig. 4-10. Graph paper is helpful in drawing practice of lower case letters.

the *waistline;* the lower is the *baseline.* Some letters, such as b and d, have extensions which go somewhat above the waistline, reaching what is known as the *cap line.* Other letters, such as j and q, have extensions which go below the baseline, reaching what is known as the *dropline.* These four lines, baseline, waistline, cap line and dropline, appear in Fig. 4-9. Alongside each lower case letter you will see arrows and numbers, indicating the sequence and direction of movement of the lines of the letters. To draw lower case letters, then, start with the horizontal guidelines shown in Fig. 4-9 and practice drawing each of the letters of the alphabet. As an alternative procedure, since you will be working with vertical letters, use graph paper as indicated in Fig. 4-10.

Both the Braddock-Rowe and the Ames lettering guides are very useful in drawing lower case letters since they can easily be used to draw the cap line, waistline, baseline and dropline correctly spaced from each other. This will help you produce lower case letters having the correct proportions.

PENCIL OR INK

The decision whether to use pencil or ink is a personal matter. If you are first learning lettering and have no drawing aids, then there is no question that working with a pencil is much easier. The same is true if you are using simple lettering guides such as the Braddock-Rowe triangle or the Ames lettering guide. With a template operated device such as Leroy or Wrico, you will find that these are geared for the use of ink.

Still, the graphite pencil remains the most popular tool for drafting work, not only for lettering, but for drawings as well. It is the easiest way, the simplest way, and lends itself more readily to erasing and corrections. This doesn't mean that pencil is automatically better. An ink line is always blacker and for lettering is certainly more readable. Ink drawings are also superior if they are to be reproduced photographically.

THINGS TO DO

- Using a sheet of graph paper, draw the numbers 1, 4 and 7. Use pencil only for this and other suggested lettering work projects.
- Draw the numbers 9, 6, 8, 3, 2, 5 and 0. For this project and the one preceding it, use Fig. 4-2 as a guide.
- Using Fig. 4-3 as a guide, draw the letters of the alphabet. Graph paper will be helpful when learning to draw letters and numbers.
- Letter a work sheet as shown in Fig. 4-4. Select a number of words and try drawing them using horizontal guide lines only.
- Using your triangles, draw a succession of slant lines and use these as guides for drawing the alphabet in slant letter form.
- Try drawing these fractions in vertical and then in slant form: 1/2, 3/4, 5/8 and 7/16.
- Try drawing the alphabet in lower case letters using horizontal guide lines only. Use Fig. 4-9 as a reference. If you find this too difficult, try drawing lower case letters using graph paper.
- Obtain graph paper and draw vertical upper case letters. Start with letters about 1/2-inch high and then gradually decrease until you can do them in a 1/8-inch size.
- Draw horizontal guide lines on blank paper and then practice drawing vertical letters. Use vertical guide lines as required.
- Using as many guide lines, vertical and horizontal as you wish, draw upper case slant letters.
- Using four horizontal guide lines and as many vertical guide lines as you wish, draw both straight and inclined lower case letters. Also draw both vertical and slant digits. Draw fractions in both vertical and slant form.

SUMMARY

- Lettering consists of capital letters, small letters and numbers.
- Lettering can be done with the help of professional lettering guides, a Braddock-Rowe triangle or an Ames lettering guide.

These are sold in art shops and are supplied with instructions. Lettering can also be done in freehand style.

- Graph paper can be used to supply vertical and horizontal guide lines when learning lettering.

Lettering may be either vertical or slanted. Lettering on drafting plates uses horizontal guide lines only or is done with the help of a lettering device.

Chapter 5
Dimensioning

Every drawing has two basic parts — lines and dimensions. The lines give you a view of the object, or may give you a number of views, just as though you were examining the object from different positions. A dimension is a description.

TYPES OF DIMENSIONS

To describe an object, we must use two types of dimensions. One of these is the size dimension and the other is the location dimension.

The two drawings in Fig. 5-1 show location and size dimensioning. Location and size are handled in the same way. If you will examine Fig. 5-1 carefully you will see that they both use extension lines, leader lines. Both require arrowheads.

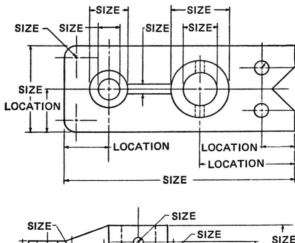

Fig. 5-1. Drawings require both size and location dimensions.

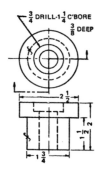

Fig. 5-2. Drawing with size dimensions only.

Because size and location on a drawing use the same drawing techniques, it is sometimes easy to confuse one with another. The purpose of a size dimension is to give the numeric value of a diameter, a width, a length or the radius of an arc. A size dimension tells you how big the object is, or how big some part of it is.

A location dimension gives the distance between different parts of the same object. For example, if the object has two holes in it, the holes may be dimensioned as 1/4 inch. This means the diameter of each hole is 1/4 inch. A dimension line marked as "two" between the center points of the holes means that the holes are 2 inches apart.

A drawing may contain only size dimensions, or both size and location dimensions. However, except for some very special purpose, it would be unlikely that a drawing would have just location dimensions. The drawing in Fig. 5-2, for example, has size dimensions only while the drawing in Fig. 5-3 has both types.

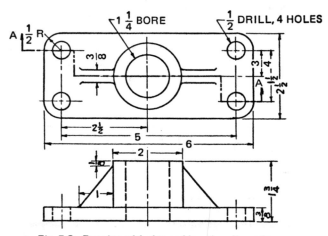

Fig. 5-3. Drawing with size and location dimensions.

There is still another type of so-called dimension, but it really isn't a dimension in the way we have defined it. A reference dimension is only used to supply additional information. To make sure that a reference dimension is not confused with size or location dimensions, we put the letters REF (an abbreviation for reference) immediately after or directly below the dimension. Unlike location and size dimensions, a reference dimension does not control the manufacture of the drawn object. Figure 5-4 shows how a reference is used.

A dimension can be used both as a size and a location dimension at the same time, provided some starting reference is supplied. The reference is simply a point or line from which measurements start. The left-hand edge or a mark near the left-hand edge of a scale is a reference, for this is generally the starting point or line for a distance measurement.

As an example, consider the drawing shown in Fig. 5-5. The left-hand edge (A) of Fig. 5-5 is regarded as the reference and measurements are made with respect to this edge. While the various dimensions given in the drawing are size dimensions, they are also distance dimensions since they are made with respect to a reference line.

In Fig. 5-5 it might seem as though one dimension is missing. While it does not appear on the drawing, this information is actually supplied. To calculate the dimension represented by the letter B, add the two given dimensions, 5/16 and 7/16, and then subtract their sum from the total dimension, 1-3/8. Of course, 5/16 + 7/16 equals 12/16. The total dimension is 1-3/8 or 1-6/16. Since 1 inch is the same as 16/16 inch, this dimension is 16/16 + 6/16 or 22/16 inches. 22/16 inches minus 12/16 equals 10/16 inch. This is the "missing" dimension.

An alternative method would have been to put in the dimension of 10/16 and to have omitted the 7/16 dimension. However, the method shown originally is preferred since it is simpler and more direct.

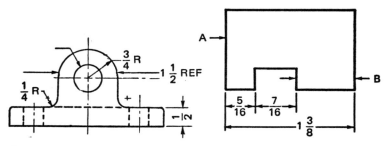

Fig. 5-4. Method of using a reference dimension

Fig. 5-5. Method for avoiding duplication of dimensions.

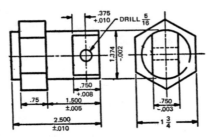

Fig. 5-6. A single dimension may be made common to two views of an object.

GENERAL RULES FOR DIMENSIONING

A good drawing is a working drawing. It is one that is complete, is based on a good understanding of the limitations of the machinery and tools used in the manufacturing process, and does not omit any information the user of the drawing will need to have.

You may not force on the person using your drawing the need for making a series of decisions or guesses. Dimensions must also be reasonable and take into consideration the kind of machinery that will be used in fabricating the object shown in your drawing.

Dimensioning is a matter of experience and common sense, both of which must be acquired. However, there are certain rules you can follow which will help you.

- If you have related views, put the dimensions between them whenever you possibly can. Figure 5-6 is an example. The view at the left is a side view; that at the right is a front view. Locate the dimension marked "1.374, – .002" and you will see it is common to both views — it is used by both.

- Do not put dimensions directly on the object you are drawing. If you do, you may find they will interfere with or will cover construction lines. Figure 5-7 shows how dimensions are positioned external to the drawn object. If the dimensions had been drawn inside the construction lines, they would have been difficult to read and certainly would have made the drawing confusing.

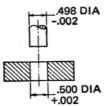

Fig. 5-7. Use extension lines to keep dimensions away from the outline of the object.

- If you have more than one view in your drawing, you will have a choice as to which you want to use for dimensioning. Use that view which shows the feature being dimensioned more clearly. This does not mean it is always possible or desirable to put all the dimensions on one view. In some cases you will have no choice. Before you start dimensioning, consider what dimensions you will need and then plan where to put them. Figure 5-8 is the drawing of a bearing. The upper drawing is a top view such that we are looking directly down on the object. The lower drawing is a side view. Note that most of the dimensioning appears in the side view.

    It may sometimes be difficult to decide which view to dimension. In that case, draw the dimensions lightly in pencil, keeping in mind that no dimension may be omitted. In the case of the drawing in Fig. 5-8, note that it would have been either difficult or impossible to put the dimensions in on the top view. Note also that the top view contains a dimension that is inside the drawing. This apparently violates the second rule given earlier. Remember, though, this is a rule and not a law. In Fig. 5-8, the inclusion of the dimension inside the object does not interfere with the construction lines.
- Never dimension a hidden edge.
- After you have dimensioned a drawing, check it carefully to make sure you haven't omitted any necessary dimension and that you haven't duplicated a dimension.
- Put small dimensions close to the object and dimensions that are larger farther away from it.

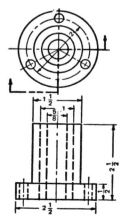

Fig. 5-8. Most of the dimensions appear in the lower drawing — the side view.

How you use dimensions may also be decided by the company employing you. Large companies that have extensive drafting departments may have a set of dimensioning style rules. The reason for this is that dimensioning can often be as individualistic as a person's handwriting. A company employing draftsmen may wish to have uniformity and standardization in its drawings. If you are preparing drawings for a personal project, obviously you can dimension for your own convenience.

Finally, as a general rule, remember that clarity is most important. Someone must read your drawing and work from it. You could, for example, cluster all your dimensions. While the drawing would be technically accurate and you would have observed every drafting rule, your drawing would be difficult to read and hence would not be professional.

Economy Of Dimensioning

Two identical drawings can supply exactly the same information, and yet one of the drawings may have more dimensioning information than actually required. Unnecessary dimensioning not only makes a drawing look more complicated, but the larger the number of dimensions the greater the chance for error. This means that as a draftsman you must do some thinking and planning before putting dimensions on a drawing.

As an example, examine the drawing shown in Fig. 5-9. Only two dimensions are supplied, but actually there are four for dimensions for A and B can be easily calculated. The radius is given as 1/2. Add this to 2 and the distance of B becomes 2 1/2. The distance A is the radius 1/2, multiplied by 2. So, 2 x 1/2 = 1, and so the distance A is 1.

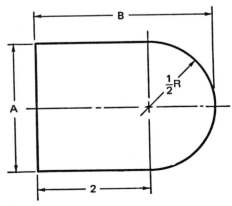

Fig. 5-9. Dimensions A and B can be calculated and so are not included here.

Do not draw any dimension more than once. Each dimension should be complete and there should be no reason for duplicating it. You might think it would be impossible to repeat a dimension, but consider that a drawing usually has a number of different views, and that each of these views may be dimensioned in some way. The very fact that you have a number of views of an object in your drawing means you must be especially careful.

There are several reasons why it is bad drafting practice to duplicate dimensions. The most important is that the duplicating dimensions may contradict each other. You may have different tolerances or the actual dimensions themselves may not agree. Any unnecessary dimensioning tends to clutter a drawing, making it more difficult to read. Consider a drawing somewhat like a telegram. Your objective should be to supply the maximum amount of information with the least number of lines and letters.

Placing Dimensions

Different techniques can be used for placing dimensions on a drawing. Which system you select will depend on your own personal taste, or it may be specified by your employer. But whichever system you use, it is important to remember that dimensions must not only be legible, but should be positioned so they do not interfere with the lines of the drawing. Dimensions can be made so they are all horizontal, or horizontal and vertical, or may even be placed at an angle.

Unidirectional Dimensioning

The technique of putting all dimensions horizontally is called unidirectional dimensioning. In this method, illustrated in Fig. 5-10, all numbers, regardless of whether the dimension lines are horizontal, vertical or slanted, are drawn horizontally. The advantage of having all dimensions drawn horizontally is that it makes it easier to read the drawing. However, this approach might also make the drawing appear more cluttered.

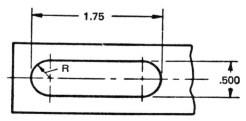

Fig. 5-10. How to place dimensions using the unidirectional system.

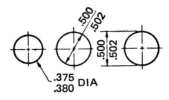

Fig. 5-11. Aligned dimensioning technique.

Aligned Dimensioning

In this drafting technique, the dimensions are always placed in line with the dimension lines, as shown in Fig. 5-11. Thus, if a dimension line is horizontal, then so are the dimensions. If a dimension line is slanted, so are the dimensions. If the dimension line is vertical, the dimensions are also inserted vertically. The advantage of this method is that the dimension line reads directly into the numbered dimension. Thus, the dimension line and its dimension are always closely associated.

In drafting, as in so many other subjects, a method or technique is often known by a number of different names. Thus, the aligned dimensioning system is also known as *rectangular dimensioning*. Figure 5-12 shows an example of the non-aligned method; that is, the dimensions appear in all directions.

Grouped Dimensions

In a particular drawing, you may have a number of dimensions following each other in the same line, forming a group. If you will examine Fig. 5-13 you will see two dimensions following each other

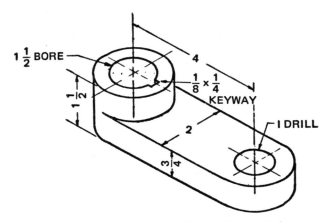

Fig. 5-12. Non-aligned dimensioning method.

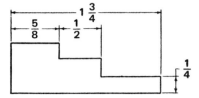

Fig. 5-13. Dimensions can form an in-line group of two or more.

directly. Note also that the "in-line" dimensions need not have been placed this way, but could have been staggered. One dimension could have been placed higher than the other (Fig. 5-14). Some manufacturers prefer *grouping dimensions* on their drawings since it does make the drawing easier to read.

Note, also, that the numbers are centered between each of the dimension lines and that the arrows of the dimension lines touch the extension lines. Where a pair of extension lines are too close to each other to permit the insertion of the dimension numbers between them, use the technique shown at the lower right in Fig. 5-13. As you can see, a dimension line is drawn between the extension lines, but the dimension itself is made to follow a line that resembles a leader. This line, however, looks like a pair of lines at right angles to each other, but not using an arrow. Note that one of the lines touches the extension line at the point where the arrow of the dimension line also touches.

Staggered Dimensions

The *staggered dimension* method means that none of the dimensions are drawn along the same horizontal line. Figure 5-14 is a drawing in which the dimensions are handled in this way. Staggering does make it a bit more difficult to read a drawing, but this technique is sometimes used to avoid crowding. This does not apply to Fig. 5-14 since it is quite obvious that the dimensions could have been grouped.

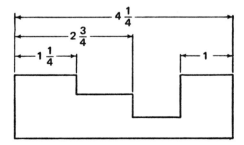

Fig. 5-14. Technique for staggering dimensions.

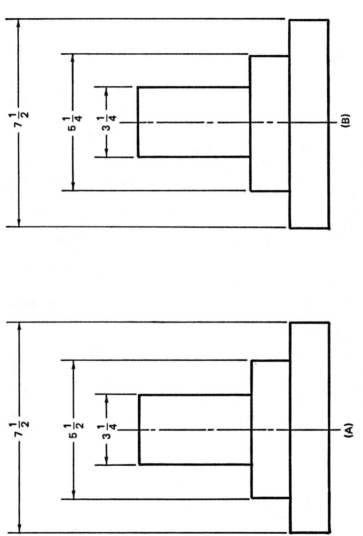

Fig. 5-15. (A). Dimensions are aligned vertically. (B). A preferable method with vertically stacked numbers offset from each other.

Stacked Dimensions

While it is good drafting practice to have a number of dimensions along the same line as shown earlier in Fig. 5-13, it isn't correct to stack numbers in a vertically aligned manner. Figure 5-15 shows two dimensioning arrangements. In Fig. 5-15A, the numbers are so arranged that they are one directly above the other. This isn't regarded as a correct technique since one number (when read in a hurry) can be mistaken for another. A preferable approach appears in Fig. 5-15B. Note that the dimensions are staggered. Also note that the smallest dimension is closest to the object and that as the dimensions increase they are spaced more and more away.

DIMENSIONING SPECIAL SHAPES

Dimensioning should not be subject to interpretation. There should be one and only one way of using dimensions in a practical application. For some shapes, exact dimensioning requires more than giving the height, width or depth of the object.

Dimensioning Angles

Figure 5-16 shows a number of angles and various ways of dimensioning them. The dimensions of an angle can be drawn inside or outside the angle. Generally, if the angle is large enough, the dimension can be placed inside. If it is very small, then there may be no choice but to put the dimension outside.

There are actually two ways of handling the dimensioning of small angles and these methods are shown in the lower part of Fig. 5-16. In the lower left drawing, the dimension is put inside the angle, while the dimension lines are placed outside it. If the angle is very small this may not be possible. Both the dimension lines and the angular dimension itself must be put outside the angle, as shown in the lower right drawing of Fig. 5-16.

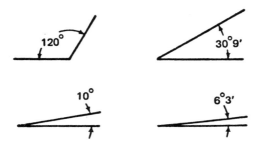

Fig. 5-16. How to dimension angles.

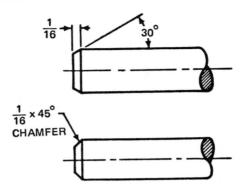

Fig. 5-17. Two techniques for dimensioning a chamfer.

Note also that although the angle itself consists of a pair of straight lines meeting at a point, the dimension lines are curved. This is a reflection of the fact that an angle is regarded as part of a circle or an arc. One further item is that the lettering of the dimension must be uniform, regardless of the size of the angle. A small angle does not mean smaller lettering.

Dimensioning Chamfers

A chamfer is an edge that has been beveled. Figure 5-17 shows two ways in which you can dimension a chamfer. In the upper drawing extension lines are used. The chamfers require two bits of information. One of these is the distance of the chamfer from start to finish. In this drawing it is 1/16 inch. The other measurement is the slope of the chamfer, shown in the upper drawing as 30° (Fig. 5-17.)

The lower drawing in Fig. 5-17 shows the alternative method of dimensioning. Unlike the upper drawing, the lower one uses a leader line drawn to the edge of the chamfer. The upper drawing is regarded as the preferred method. The dimensioning technique in the lower drawing is used when necessary to simplify the drawing.

Dimensioning The Radius Of An Arc Or Circle

The radius of a circle is the shortest line that can be drawn from the center of the circle to any point on the circumference. A radius is equal to one-half a diameter.

A radius can be used in connection with a circle, or with an arc, since an arc is a part of a circle. Whenever you draw an arc or a circle and want to show the radius, you can do so by drawing a dimension line, and then including the dimension and the letter R as shown in Fig. 5-18.

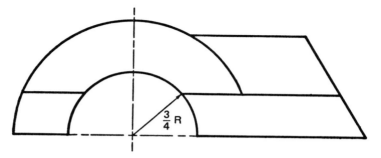

Fig. 5-18. Technique for dimensioning the radius of a circle or arc.

Sometimes you may have drawings on which you will be required to have a large number of arcs or circles. Figure 5-19 shows how such drawings are commonly dimensioned. The leaders are not only curved but have the same arc of radius as the arcs or circles in the drawing. This makes the drawing look more professional. You will also sometimes find angular dimensioning used in conjunction with rectangular dimensioning.

You can dimension a circle either inside or outside, depending, of course, on the size of the circle and the closeness of adjacent dimensioning. There is no point to cramping lettering inside a circle if the circle is small and there is ample room outside it. When dimensioning outside a circle, you can use either extension lines or leaders. Figure

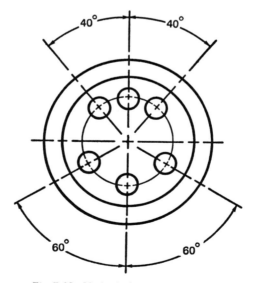

Fig. 5-19. Method of dimensioning arcs.

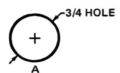

Fig. 5-20. (A). How leaders are used outside a circle. (B). The dimension line is inside the circle. (C). The dimension is outside the circle with a pair of extension lines going to the diameter of the circle from the dimension line.

5-20A shows how leaders are used outside a circle. The leaders should be the extension of an imaginary diameter of the circle. A diameter is any straight line touching the circumference at two points and passing through the center of the circle. To produce the type of dimensioning shown in Fig. 5-20A, draw a diameter lightly through the circle but continue it for a short distance beyond the circumference. Then draw the arrows as shown and letter in the dimension.

The dimension can be just a fraction, such as 3/4 or 1/2, or may consist of a dimension followed by an explanatory note. The dimension can appear as "3/4 through hole," or as "3/4 hole," or as "3/4 hole countersink 1/8."

In Fig. 5-20B the dimension line is inside the circle and is a diameter of that circle. In Fig. 5-20C the dimension is outside the circle with a pair of extension lines going to the diameter of the circle from the dimension line.

Dimensioning Multiple Holes

If an object has two or more identical holes with all the holes to be handled in the same way during manufacturing, you can achieve economy of dimensioning by using the method shown in Fig. 5-21. While all the required holes must appear in the drawing, just a single dimension is needed, which can then be applied by the user of the drawing to all the holes. Not shown in Fig. 5-21 are the necessary positioning dimensions, always supplied in a complete drawing.

DIMENSION LAYOUT

If you want to see just what your drawing will look like without spoiling your efforts, put a sheet of tissue paper (called a tissue overlay) over your drawing and then letter in, rough, the various dimensions. Use a soft pencil and erase as you need to do so. When you are satisfied that the dimensioning is satisfactory, or it follows required practice, remove the overlay and use it as your guide. An overlay is just a sheet of thin paper, smooth and transparent, that lets you see the drawing easily, but protects it from your dimensioning sketch. You can fasten the overlay to the drawing by bending the overlay so that

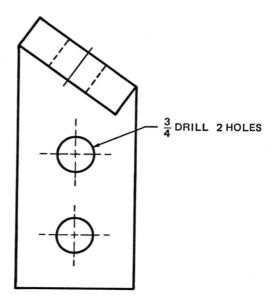

$\frac{3}{4}$ DRILL 2 HOLES

Fig. 5-21. Method for dimensioning holes of identical sizes.

a small section of it folds over the top of the drawing. Fasten this rear section of overlay to the drawing with a small bit of removable tape.

Whenever possible, do not cross lines used in dimensioning. This is one of the advantages of using a tissue overlay. It will help you plan the placement of dimension lines. You will also find it helpful to put the dimension line for the shortest length nearest the outline of the object. Insert dimension lines in the order of size. Following this procedure, the longest dimension will be the outermost one.

Obviously, you may not omit a dimension to make your drawing look less crowded. While you should try to have as professional looking drawing as possible, the important thing isn't your drawing but the construction or manufacturing that can be done with its help. If there is no possible way in which you can avoid crossing lines, be sure to make a break in the extension line or the leader line at the point of crossing.

Size Of Dimension Numbers

The size of numbers used in dimensions is a matter of personal preference or may be a requirement of a drafting department. However, if a particular size is selected it should be uniform throughout the drawing. Going from one number size to another in the same drawing not only spoils the appearance of the drawing but is actually more work.

True Dimensions

Sometimes, in a drawing, you will have a view that is shortened, or which will show an object in perspective. While such views may seem more "real" to us because that is how we actually see objects, such views do not represent the true dimensions of the object. Whenever you dimension, be sure to select a view which shows the true length, width or depth. As a general guide, you might consider putting as many dimensions as you can on the front view. To follow this suggestion, like so many others, does require some common sense. The reason for selecting the front view is that it often shows most of the characteristics of the object. There are enough exceptions to this statement, though, to make you cautious about accepting it without reservation.

Notes

Sometimes you will need to add notes to your drawings to supply information concerning certain features of the object. There are two types of notes, as shown in Fig. 5-22. These are specific notes and general notes.

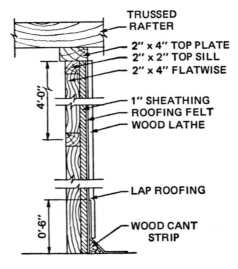

Fig. 5-22. Specific notes appear alongside the drawing. General notes appear below it.

Still referring to Fig. 5-22, you will see that this has a number of specific notes. These are placed adjacent to the drawing and connected to it by leader lines. The general notes are put beneath the drawing and are headed by the word "NOTE". This word may be included or omitted, depending on the drawing style being followed. Figure 5-22, incidentally, is a type used in building construction work.

A specific note, as its name implies, describes only a certain part of the object and applies only to that part. A general note has a meaning completely opposite that of the specific note. A general note applies to all parts of the object. As an example, if the complete object is to receive some kind of finish, this information could be supplied in a general note. On the other hand, if some part of the object is to receive special treatment, such as milling, grinding or deburring, this could be covered in a specific note pointing directly to that part of the object scheduled to receive such treatment.

You can letter general notes directly beneath the object you have drawn or you can put such notes in the title block. The title block is generally an outlined rectangle in the lower right-hand corner of the drawing sheet. The title block, shown in Fig. 5-23, follows this general form, although it may vary from one drafting company to another. It usually has provision for the name of the draftsman, the date, the name of the drawn object and the project number. It may include space for the initials of a checker, the sheet drawing number. The drawing may just be one of a large number of drawings and the scale that is used, if any.

One of the advantages of specific notes in machine drawings is they can help eliminate numerical dimensions. Under such circumstances the specific note is referred to as a dimensional note. A dimensional note, for example, might read, "1/2 drill, two holes." In this example, without the help of the dimensional note, you

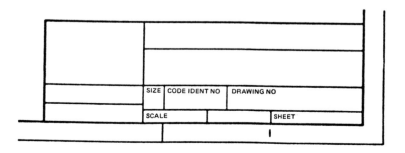

Fig. 5-23. Title block.

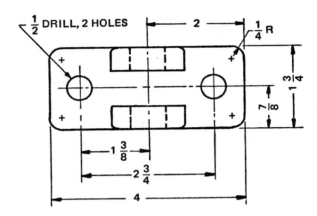

Fig. 5-24. Specific notes help minimize the need for showing dimensions.

would need to indicate numerical dimensions in two places in the drawing. Figure 5-24 shows the use of this drafting technique. Both specific notes and dimensional notes require the use of leaders, but general notes do not.

Dos and Don'ts Of Dimensioning

- Plan all dimensions on a piece of scrap or on a tissue overlay. Do not proceed immediately to dimensioning after completing a drawing without thinking about the positioning of the dimension lines with respect to each other and the possibility of eliminating unnecessary dimensioning.
- If your drawing has more than one view, make sure you do not duplicate dimensions in the various views.
- Always put the shortest dimension closest to the drawing, and the larger dimensions further away, in order of size.
- It is permissible to have extension lines cross each other. If it is necessary to do this, make sure the extension lines are continuous, unbroken lines.
- Extension and dimension lines should have a lighter weight of line than the lines used in drawing the object. The user of the drawing should never be able to confuse object lines with extension or dimension lines.
- While extension lines may cross, try to avoid crossing dimension lines.
- Try to line up dimensions horizontally whenever you can. If you have a choice between staggering dimensions or lining them up horizontally, use the horizontal arrangement if possible.

- While inch marks (") are generally omitted from drawings, use them when necessary to avoid confusion. There is a big difference between 2 holes and 2" holes.
- If you have a group of dimensions in a single horizontal line it is common practice to omit one of the dimensions.
- Add notes to dimensioning as required. Such notes indicate some aspect of the manufacturing process — 5/8 drill or .664-666 ream.
- When drawing the radius of an arc or circle always include the letter R. Write the letter D when drawing a diameter.
- Never use an object line as a substitute for a dimension line or an extension line.
- You can do dimensioning inside or outside the drawn object. Generally, it is better drafting practice to put dimensioning on the outside.
- You can use either common fractions or decimals or both on a drawing. The modern trend is more toward the use of decimals. Thus, instead of lettering 1/4, letter .25 instead. Instead of 1/8, use .125 instead. When lettering decimals it is necessary to include the decimal point.
- Since the purpose of a drawing is to enable someone to manufacture an object, perform an assembly, construct a building, or work on a project of your own, it is important for dimensions to be not only legible, but to be positioned so they do not interfere with the lines of the drawing.

TOLERANCES

When indicating dimensions on a drawing, you will also be required to show the permissible *tolerances*. If, for example, the object to be manufactured is to have a through hole — a hole which goes completely through the object — you must also include information that will tell the machinist how much larger or how much smaller he may make that hole. A tolerance represents the outer limits in size. Naturally, a tolerance must take into consideration the machine that is being used to manufacture the part shown in the drawing. It is generally useless to specify a tolerance of a thousandth of an inch if the machinery available for manufacturing is a coarse type that can only work with tolerances of hundredths of an inch.

A tolerance may be plus, minus or both. If, for example, you want a hole of 1/2-inch diameter drilled through a piece of metal but do not want the hole to be smaller than this diameter, you would simply show the diameter with a plus tolerance. If the maximum diameter is to be 1/2-inch, but never more than this at any time, you would probably show a minus or negative tolerance. If the hole diameter could be either larger or smaller than 1/2-inch, then your dimensioning would include a plus-minus tolerance.

Figure 5-25 shows a number of different ways of indicating tolerances on drawings. Figure 5-25A shows a negative tolerance for the dimension at the top and a plus tolerance for the dimension at the

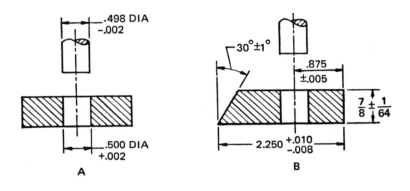

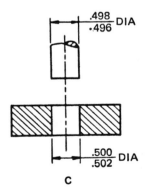

Fig. 5-25. (A). Negative tolerance for the dimension at the top and a plus tolerance for the dimension at the bottom. **(B).** Several ways of indicating plus and minus tolerances. **(C).** Two numbers are given in this method of indicating tolerance.

bottom. According to the drawing, the maximum dimension of the rod is to be 0.498 inch, but it may be as much as 0.002 inch smaller. Subtract 0.002 from 0.498 and the result is 0.496. This number is the dimension limit. Thus, the rod in the machining process may have any dimension from 0.496 inch to 0.498 inch. If, after manufacture, the rod was calipered and found to have a dimension of 0.497 inch, then it would be satisfactory since this is within the tolerance limits of 0.496 and 0.498. A rod measuring 0.500 inch, though, would be rejected.

Figure 5-25B shows several ways of indicating plus and minus tolerances. If the plus-minus tolerances are the same, the symbol used is a plus sign with a minus sign drawn beneath it. If the tolerances are not the same, they are shown separately, with the plus sign in front of one number and the minus sign in front of the other. Examples of both are shown in Fig. 5-25B.

A common error in drawing plus-minus symbols is to make the minus sign look like a dash. The minus sign should have the same length as the horizontal line of the plus sign. And the horizontal and vertical lines of the plus sign should have identical lengths.

Still another way of indicating tolerances is shown in Fig. 5-25C. In this method, two numbers are given. These two numbers represent the upper and lower limits of the dimensions. The advantage of this technique is that it minimizes possible arithmetic errors on the part of the person using the drawing. It is easy enough to add or subtract a tolerance number from the main dimension, but it is also easy to make a mistake in doing so.

You will often see the number zero preceding a number in a text such as this. The purpose is to emphasize the position of the decimal point. Typical examples are 0.007, 0.09, and 0.1085. However, the zero to the left of the decimal point is not included in drawings. These numbers would be shown as .007, .09, and .1085 on a drawing. The inch symbol (″) is not used. It is understood that tolerances are in fractions of an inch.

Size Marks

Inch marks (″) and feet marks (′) are generally omitted from drawings when the size is obvious. In drawing a small object, for example, inch marks will not be included. If a drawing needs to use both feet and inches, a dash mark is always included. For example, a dimension that is 2 feet, 1 inch, would be drawn as 2′-1″.

The radius of an arc or circle does not use degree marks, but degrees and minutes are used when dimensioning an angle. Thus, an

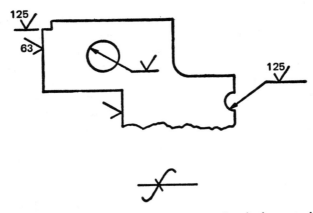

Fig. 5-26. Various types of finish symbols. The numbers in the upper drawing are codes to indicate different required finishes.

angle might be dimensioned as 25° or as 31°5′. The symbol (°) represents degrees and the symbol (′) means minutes. A minute is the sixtieth part of a degree. Use the letter R to indicate a radius and the letter D to indicate a diameter. These are sometimes omitted when it is obvious that a radius or a diameter are intended. When a circle appears on a drawing its dimension is sometimes shown by the size of the drill needed to produce it. Thus, a 3/4 drill is capable of producing a hole having a diameter of 3/4″.

Finish Marks

There are a number of ways of finishing the surface of an object. It may be plated, polished, milled, etc. Various symbols used to indicate that a surface is to be finished are shown in Fig. 5-26. One of these has a resemblance to the arrowhead, but it is not an arrowhead and do not draw it as such. It is more like the letter V. The V symbol touches the surface to be finished but does not go through it. The other finish symbol resembles a letter f that has been tilted. It goes through the visible outline and is symmetrical on either side of it.

THINGS TO DO

Use whatever tools you may require for these drawings, including triangles, T-square and protractor. Do the drawings in pencil and then ink in.

- The face of an object is a rectangle measuring 12″ x 16″. Make a one-third scale drawing and dimension both edges.

- Draw three angles. The first is 9°, the second is 30° and the third is 55°. Dimension each of these angles.
- A chamfer has a length of 1/8 inch and an angle of 45°. Make a drawing of this chamfer and dimension it.
- The object shown in Fig. 5-27 is a metallic ring with a hole drilled through its center. The ring has an outside diameter of 6 inches. The center hole has a radius of 3 inches. The ring has a thickness of 2 inches. Draw this object and dimension it.
- The object shown in Fig. 5-28 has a taper of 12°. The through center hole has a radius of 3/4 inch. Duplicate this drawing and show these dimensions.
- Draw three circles having diameters of 1/2 inch, 1 inch and 2 inches. Use three different methods of dimensioning them.
- The drawing in Fig. 5-29 is a hexagon. Each side has a length of 4 inches. The distance from A to B is 6.9 inches. Each interior angle of the hexagon is 60°. Draw this unit and dimension it. The vertical and horizontal centering lines are drawn incorrectly. Correct them in your drawing.
- Draw a pair of concentric circles. The inner circle is to have a radius of 1-1/4 inches while the outer circle should have a radius of 2-3/4 inches. Dimension both circles.
- In drawing Fig. 5-30, side A has a dimension of 4 inches. Side B has a length of 2-3/4 inches and C has a length of 1-7/8 inches. Draw this object and dimension it. The maximum outside diameter of the circle shown in the front view is 8 inches.

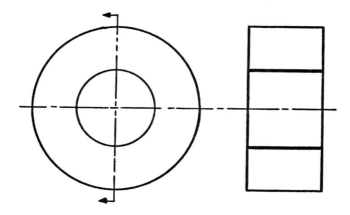

Fig. 5-27. Metallic ring with hole drilled through center. Cutting plane divides object into two parts.

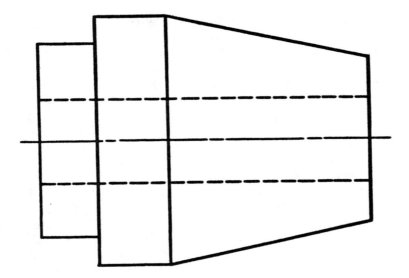

Fig. 5-28. Dimension the taper.

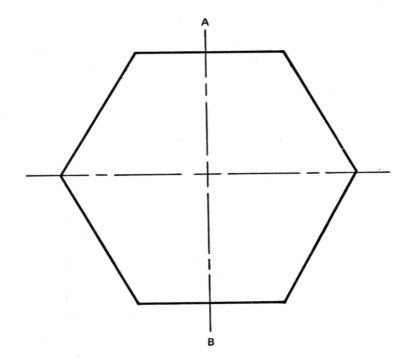

Fig. 5-29. A hexagon is a geometric figure with six sides.

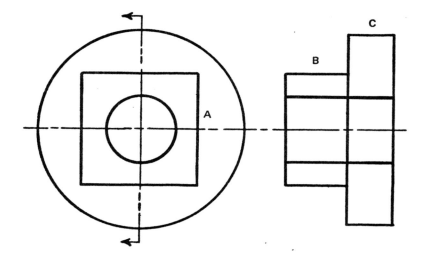

Fig. 5-30. A problem in dimensioning.

SUMMARY

- Drawings can have two types of dimensions – size and location. A location dimension gives the distance between different parts of the same object. A reference is a point or line from which measurements start.
- Unnecessary dimensioning makes a drawing look more complicated. Do not draw any dimension more than once. The technique of putting all dimensions horizontally is called unidirectional dimensioning. In aligned dimensioning the dimensions are always placed in line with the dimension lines. In a non-aligned method the dimensions appear in all directions. Staggered dimensions mean that none of the dimensions are drawn along the same horizontal line. Dimensions are sometimes stacked in a vertically aligned manner.
- The dimensions of an angle can be drawn inside or outside the angle. Circles can be dimensioned inside or outside the circle.
- Dimensioning numbers should be uniform in size throughout the drawing.
- Views that are to be dimensioned should be true views that show the true dimensions of the object being drawn.

- A specific note describes a certain part of an object. A general note applies to all its parts.
- A tolerance can be plus or minus or both. Tolerances are in fractions of an inch.
- Finish marks are used to indicate the kind of finish the surface of an object is to have.

Chapter 6
Projections

Although an object can be examined in many different ways, it is quite customary in drafting to examine it from the top, front and side. The problem in drafting is that we must draw a three-dimensional object on a flat surface, or a surface having two dimensions only.

In Fig. 6-1 we have the drawing of a machined part. Even though you may never have seen this object before, you have no problem in looking at this drawing and then recognizing the actual object from the drawing. You can see the object has holes in it and that these holes are at various angles to each other. You have some idea of the thickness of the object. Further, instead of three or four views, you have just one view. With that single view you are able to absorb a considerable amount of information about the object. Figure 6-1 is drawn in perspective. While it is a two-dimensional drawing, the object does look as though drawn in three dimensions. It appears to have width, length and height.

The trouble with such a drawing is that we cannot really dimension it. As an exaggerated example of what we mean, consider looking along a pair of railroad tracks. Imagine you are standing between the two tracks looking at them as they go off into the distance. If

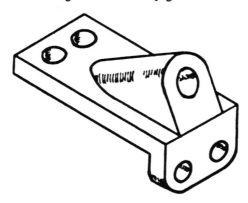

Fig. 6-1. Perspective drawing of a machined part.

the roadbed is flat, it will seem to you that the rails come together. In the far distance it may even appear as though they do touch. But you know that the rails must always be equally distant from each other. If you were to draw what you see and were to make a sketch of the rails, you would need to picture them as coming closer and closer. It might make an attractive drawing and it might even look realistic. But you could not use such a drawing for the manufacture of a rail roadbed.

For this purpose you would need to draw one or more views to which you would add location and distance measurements. Oddly enough, your drawing might not look as real as the one in which the tracks seemed to come together. What we look at and think of as real from a measurement point of view may not be real at all.

CENTRAL PROJECTION

Suppose you have a rectangular piece of material, such as stiff cardboard, and that you have connected four exactly equal lengths of string to it, one string to each corner. You can tie all the loose ends of the strings together, until they come to a point, as shown in Fig. 6-2. If you keep the four lengths of string quite tight, then you could connect a succession of smaller and smaller rectangles to the strings.

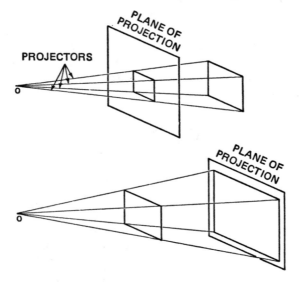

Fig. 6-2. Examples of perspective or central projection. The projectors meet at a common point. The plane of projection is an imaginary surface on which we have a flat picture or a view of one or more surfaces of an object.

The type of projection shown in Fig. 6-2 (there are other kinds as well) is called *central projection* because all the projectors or lines come together at a single point — the letter O at the left in the drawings. Instead of strings, imagine your eye is at point O at the left. Replacing the strings we now have "lines of sight," but the effect is the same. Theoretically, as these lines extend further out into space, the rectangle becomes larger. In the upper drawing of Fig. 6-2 our cardboard represents a plane surface; that is, an object having only two dimensions — length and height — but no thickness or width.

Also in the upper drawing of Fig. 6-2 we have inserted an imaginary surface between point O at the left and the rectangular bit of cardboard at the far right. The lines extending from each corner of the cardboard to point O pass right through our imaginary surface, which we call a *plane of projection.* You can think of the plane of projection as a sheet of glass if you like, with the lines of projection able to pass through it without any trouble. Now imagine we are able to put a tiny pencil dot right at the point where each line of projection passes through the glass. We will then have four such points. If we connect them with pencil lines, we will have a picture of the front of the cardboard. Note in the upper drawing of Fig. 6-2 the picture on the plane of projection is smaller than the actual surface area of the cardboard. In the lower drawing of Fig. 6-2 the picture we get on the plane of projection is larger than the surface area.

This increase or decrease in the size of the projected figure raises a problem for we don't know the true size. The only way in which we could get the true size using this technique would be to superimpose the plane of projection directly on the face of the object. In that case we would be defeating our purpose for we would no longer have a plane of projection. While central projection is interesting and is helpful in certain kinds of free-hand drawings, it isn't of value in drafting.

PARALLEL PROJECTION

Look once again at the lower drawing in Fig. 6-2 and you will see that the object and its image on the plane of projection have different sizes. If, however, we do not insist on having the projectors come to a point, but arrange them parallel to each other as shown in Fig. 6-3, the object and its projection will have the same size. There are two requirements, though. The projector lines must indeed be parallel to each other. Furthermore, the object and its projection must also be parallel to each other.

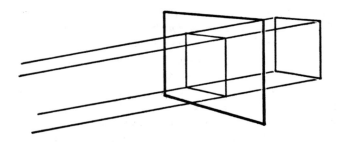

Fig. 6-3. Example of parallel projection. The object and its projection have the same dimensions. The object shown in this drawing could be one face of a box.

Parallel projection is known as orthographic projection or, less commonly, orthogonal or right-angle projection. With this type of projection, the projectors are perpendicular both to the object and to its projection.

We rarely encounter simple objects alone, such as boxes, in drafting work. Some geometric shapes are rather complex, while others consist of combinations of simpler shapes. The object in Fig. 6-4 isn't a box, but we can consider it as a pair of boxes joined to each other. However, to project the front view of this object, we need more projectors. This time we have six of them. The procedure, though, remains the same. The projectors are all parallel to each other, while the plane of projection is parallel to the front face of the object. The numbered points on the plane of projection correspond to similarly numbered points on the object to be drawn. However, after we connect all the points on the plane of projection with straight lines, all we have will be a drawing of the front of the object. We still do not know, by looking at the drawing on the plane of projection, how deep or thick the object is.

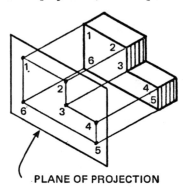

PLANE OF PROJECTION

Fig. 6-4. A more complex object than a box will require a larger number of projectors.

THE THREE VIEWS

With orthographic projection our technique is to enclose the object with planes that are parallel to the various surfaces of the object. We then imagine perpendicular lines coming out of the object, piercing the planes. By connecting each pierced point with lines, we get projected views which have true dimensions.

Since the objects we are going to draw are three-dimensional, we can get more data about them by having three planes of projection, or three views. One of these would be the front; the other two would be the side and top views or planes.

The Front View

The projection in Fig. 6-4 is called a front view. The drawing we get on the plane of projection is just as though we held the object right in front of our eyes, looking neither to the right or the left. By itself, the front view does give us some information about the object, but quite often the front view isn't enough to let us visualize the item it represents. As an example, imagine we have a right cylinder behind the plane of projection. The front view of this cylinder would appear as a rectangle. With a sphere, such as a baseball, the front view would be a circle.

Side And Top Views

The technique for producing side and top views is the same as that used for the front view. We draw parallel projector lines from the side and the top and let these lines intersect or pierce the plane of projection. We can use the same object illustrated in Fig. 6-4. Draw these additional views as shown in Fig. 6-5.

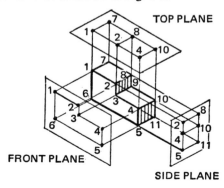

Fig. 6-5. Three views of an object projected onto front, side and top planes of projection. These are called the front, side and top planes or views.

Ordinarily, we do not number the points on the object or those on the projection, but have done so in Fig. 6-5 to show you the relationship between the various points in the planes of projection and corresponding points on the object. At this time there are two important points you should keep in mind. The first is that the three planes of projection are each parallel to some face of the object. And the second is that the lines of projection are perpendicular both to the object and to the planes of projection. If you were to measure the angle that each of the projection lines makes with the object and the planes of projection, you would find them to be right angles.

Figure 6-5 is for purposes of explanation only, and you will note that the object is turned to one side. Usually we have a front view so arranged that we look at its front view "head on," with the front view facing us directly. When we do this, we look at the object "head on," as though it were being held directly in front of our eyes, just a few inches away.

The side and top views follow the same approach. We look at one side of the object and also down on it. In Fig. 6-5, because the object is turned to one side, it is easy for us to see it in its entirety. However, a more common approach is shown in Fig. 6-6. The drawing at the lower left is the front view; the one to its right is the side view, while the remaining drawing is the top view.

If you will now go back to Fig. 6-5 for a moment, you will see how we obtained these three views. We have projectors coming out of the corners of the object. If you will examine the front of the object, you will see that each corner is numbered and that we have projector lines coming out of those corners.. These projector lines are perpendicular to the object, and so form right angles with it.

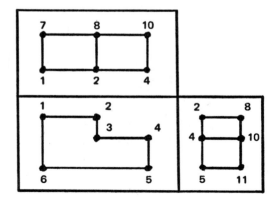

Fig. 6-6. Front, side and top views in the same plane.

In front of the object we have a front plane. Imagine this as a stiff sheet of very thin paper placed absolutely parallel to the front of the object. Wherever a projector comes through the paper it will produce a point which we can identify with a number. After connecting all the points, we will get the front view, as shown in the lower left in Fig. 6-6.

We can produce side and top views in the same way. Just imagine a sheet of paper placed perfectly parallel to the selected side and projectors coming from the object and piercing the paper. Again we number the points. When we connect them with straight lines, we will get the side view or side projection as shown to the right in Fig. 6-6. Similarly, by putting a sheet of paper parallel to the top, we can use projectors to get the top view.

Now let's go back to Fig. 6-5 just for a moment and examine the front plane. The front plane consists of lines connecting points 1, 2, 3, 4, 5 and 6. But why just these points? Those are the only points we could actually see if the object were held right in front of our eyes. We could not, for example, see points 7 and 8; nor could we see points 10 and 11. If you will examine the drawing at the lower left in Fig. 6-6, you will see that we have only points 1 to 6 inclusive. By connecting each of these numbered points, we obtain what is known as the front view of the object. The front view does not show points 7, 8 and 10 for it is impossible for us to look through the solid object.

The Importance Of Three Views

One of the problems in drafting is that you must be able to visualize an object from the three views that are supplied — the front, side and top projections. The difficulty is that each of the views is just two-dimensional and you are being asked to imagine a three-dimensional object given two-dimensional material. The solution would be to hand you the object and let you look at it, but this isn't always practical or possible. You may be required to draw an object, or some part of it, that simply cannot be supplied. It is either too large, too heavy, or for some reason may be physically impossible to obtain or see.

As an example, consider the three views shown in Fig. 6-7. Without looking at the drawing, try to visualize what sort of object is represented by these views. If you will now look at Fig. 6-8, you will see that the three views are of a rectangularly shaped box. As a simple exercise, compare the numbers on the views with those on the object.

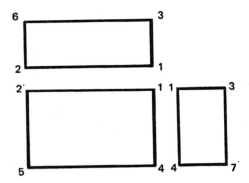

Fig. 6-7. The three views of an orthographic projection.

The box illustrated in Fig. 6-7 has no holes drilled in it, no cut corners, no bevels, no mortising or routing. But now consider the next drawing, the orthographic projection of Fig. 6-9. Can you visualize the object? Move ahead to Fig. 6-10 and you will see what it looks like. Now compare the orthographic projection with the drawing of the object. Pay particular attention to the side and top views to understand why each of the lines in the orthographic projection was necessary. An excellent way of learning the significance of each line would be to take the drawing of the object, and then, with this book closed, try to make your own orthographic projection. When you are finished, compare what you have drawn with Fig. 6-9.

As a further exercise, there are three orthographic projections appearing in Fig. 6-11. The objects representing these projections are shown in Fig. 6-12A through 6-12C.

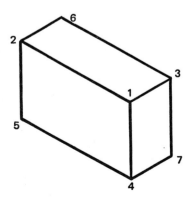

Fig. 6-8. The three views shown in Fig. 6-7 represent this rectangularly shaped box.

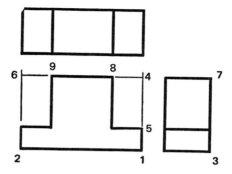

Fig. 6-9. Determine the object represented by this orthographic projection.

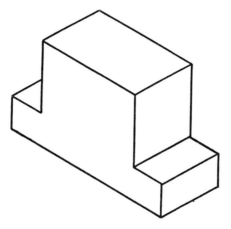

Fig. 6-10. This is the object shown in orthographic projection in Fig. 6-9.

You should note that Figs. 6-11A and 6-11B are numbered. This technique follows the practice shown earlier in Fig. 6-4. However, the reason for numbering the corners was to help you identify projection lines. In regular drafting practice these numbers are always omitted and so Fig. 6-11C doesn't use them. As a further example, go back to Fig. 6-6 and then compare it with Fig. 6-13. These are the same drawings, but Fig. 6-13 is cleaner looking and is more professional.

ADDITIONAL VIEWS IN ORTHOGRAPHIC PROJECTION

Ordinarily, in orthographic projection, just three views are enough to supply sufficient information about an object, enough data so the object can be constructed. Less often, additional views are used,

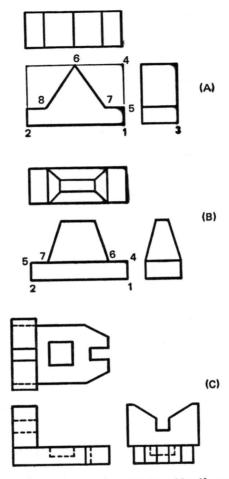

Fig. 6-11. (A). Corners are numbered to help you identify projection lines. (B). Another numbered drawing. (C). Numbers are not used here.

such as a left side, rear and bottom views. These additional views, plus the usual top and right side views, are illustrated in Fig. 6-14.

You can consider these six views as projections on transparent flat surfaces, as shown in Fig. 6-15. If you will imagine each edge of the box as hinged, then the box can be opened so it is flat. When this is done you will have the drawing of Fig. 6-14. The same object is used in Figs. 6-14 and 6-15. Figure 6-14, however, is easier to draw, and is much more practical from a drafting point of view. With Fig. 6-14 you must be able to visualize the object, something that isn't necessary with the drawing of Fig. 6-15.

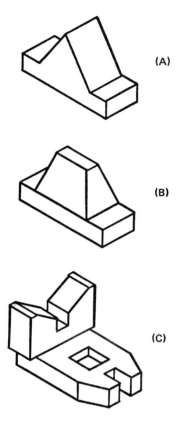

(A)

(B)

(C)

Fig. 6-12. (A). Object corresponding to the projection in Fig. 6-11A. **(B).** Object corresponding to the projection in Fig. 6-11B. **(C).** Object corresponding to the projection in Fig. 6-11C.

OBLIQUE PROJECTION

While orthographic projection is widely used in drafting, it does have its limitations. In Fig. 6-16 we have, instead of a sheet of cardboard, a box at the right. Extending from each of its corners we have lines of projection.

In the lower part of the drawing we have perpendicular projectors. These projectors pass through a plane of projection. Imagine this plane of projection as a sheet of very thin paper. When the projectors pass through this paper we can imagine they produce tiny points. In the case of orthographic projection, by connecting these points with straight lines, we get a reproduction of the side of the box facing the plane. But that is all we will see. If we do not know there is a box

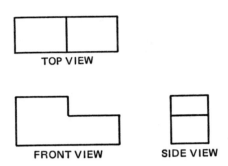

Fig. 6-13. Simplification of the three views of Fig. 6-6.

behind the paper, the drawing we have made on the plane of projection isn't going to be of much help in visualizing the box. All we will know about the object behind the plane of projection is that one of its surfaces is rectangular.

If we now adjust our projectors so they are no longer perpendicular to the box, but form some angle other than 90°, we will have the situation shown in the upper part of Fig. 6-16. The projectors are still parallel to each other, but are now tilted upward. As a result,

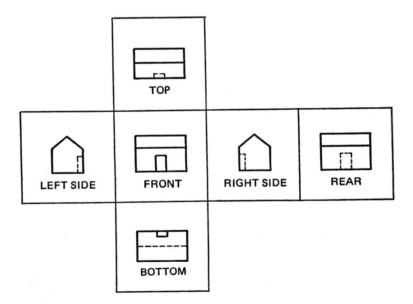

Fig. 6-14. Orthographic projection using six views.

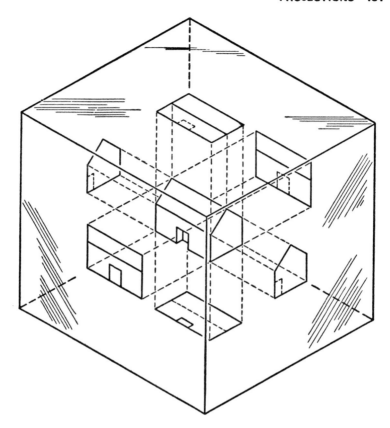

Fig. 6-15. With the object enclosed on all sides, lines of projection produce six views. The object is in the center of the box.

the projection of the box onto the plane of projection now gives us two views — the front of the box and the top.

There is something very important you should note at this point. The box is a three-dimensional object. It has width, length and depth. The plane of projection, however, is only two-dimensional for theoretically it has no thickness. As a result, whatever views we show on the plane of projection are also two-dimensional. Consequently, the front and the top of the box in the *oblique projection* are as shown. However, there is a notable difference between orthographic and oblique projection. The orthographic reveals only two dimensions in this drawing, length and height. The oblique projection reveals three dimensions — length, height and depth — but does so in two-dimensional form. We may have trouble in trying to imagine what sort of shape our object has.

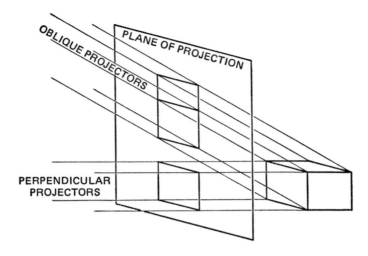

Fig. 6-16. Orthographic projection in the lower part of the drawing reveals just the front of the object. Oblique projection in the upper part of the drawing shows top and front of object.

ISOMETRIC DRAWING

An orthographic drawing can describe accurately the shape and size of an object and even reveal details of its interior. It is often difficult to visualize the object even though you may be supplied with three orthographic views plus a supplementary or auxiliary view. Further, the details of an object may be so complex that sometimes the orthographic projection with all its different views becomes a jumble of lines. For the professional draftsman with years of experience, reading such a drawing is easy. But for those who do not have the benefit of long experience, an orthographic drawing can be exasperating.

Since the purpose of a drawing is to supply manufacturing information *plus* understanding, another form of drawing is sometimes used. Known as *isometric drawing,* it is a pictorial method that resembles perspective drawing, but with its principal dimensions drawn to scale.

Pictorial drawings are three-dimensional types. If you draw a picture of a house, box or any other object, even though you are limited to the two dimensions of a sheet of paper, you can manage to create the illusion of three dimensions.

An isometric drawing borrows from perspective and from orthographic. It avoids the limitations of orthographic and of perspective, but the principal dimensions of an isometric drawing are drawn to scale.

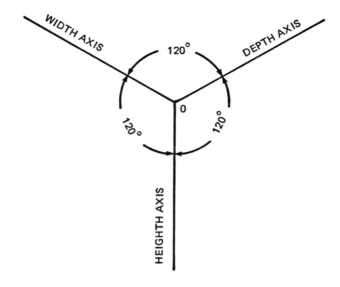

Fig. 6-17. The three axes of an isometric drawing.

Isometric Axes

The word isometric means equal measure, with the dimensions of length, width and height each represented by an axis, as shown in Fig. 6-17. The three coordinate axes are drawn 120° apart.

The three axes represent true lengths and so we can make measurements along these lines. We can also make measurements along any lines which are parallel to these three axes. A line, parallel to an isometric axis, is known as an isometric line. Such a line is drawn in its actual length and can therefore be measured. However, any line which is *not* parallel to an axis or to an isometric line is not shown in its true length. It may be either longer or shorter than true length, and is called a non-isometric line.

Using these three we can construct a cube as in Fig. 6-18. We have a circle around each of the axes in this drawing, but the only purpose of this circle is to help identify the axes for you. In this illustration we have the cube in four different positions, but actually we can start a drawing at any corner where three mutually perpendicular edges meet.

Examine the three axes in the left drawing of Fig. 6-18 and imagine you have a pin stuck in at the point where these axes meet. Also imagine that the drawing can pivot around this pin. If you rotate the drawing, you will get the succession of other drawings shown in Fig. 6-18.

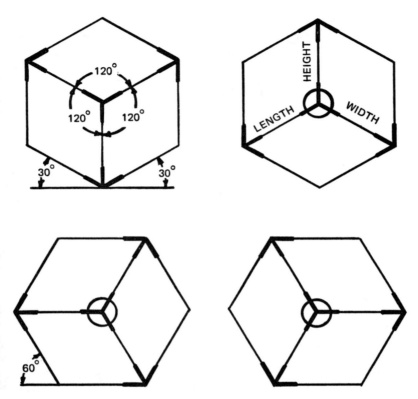

Fig. 6-18. In each of these drawings a circle is used to identify the three isometric axes. This circle would not appear in a drawing.

In the usual perspective drawing of a cube, the angle between the lines representing length and height would be a right angle. Similarly, the angles between the lines marked height, length and width would also be 90°. But in this isometric drawing, these angles are 120°. Please note that Fig. 6-18 is an isometric drawing.

We did not obtain these cubes by projection, but by drawing them directly. We did this by first setting up the length, width and height axes and then constructing our object directly on these axes. The dimensions shown in such a drawing will be true dimensions.

Practice Exercise

It would be helpful to draw isometric axes. To do so, draw a horizontal line on a sheet of paper, and then locate a point somewhere along its center. Measure an angle of 30° and draw a line with a downward slope so that it ends at the center point. Do this left

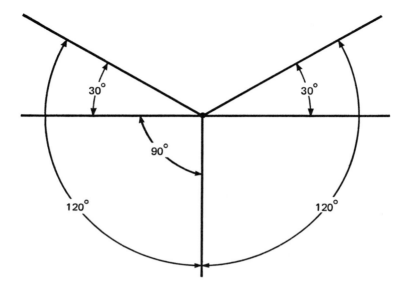

Fig. 6-19. To make isometric axes, draw a pair of 30° angles, using a 30–60–90 triangle. Drop a perpendicular from the junction of the two slant lines where they meet at the horizontal lines.

and right of the center point and you will have a pair of lines as shown near the bottom of the left in Fig. 6-18. Now drop a perpendicular from the junction of the two 30° lines. You will have three lines with angles of 120° to each other, as shown in Fig. 6-19.

You can continue and construct an isometric box by drawing lines parallel to the three axes and having the same lengths. Connect all lines and the result will be an isometric box.

The Need For Isometric Drawings

In Fig. 6-20 we have three drawings. The top and front views of an object are drawn in orthographic projection while the bottom drawing is isometric. All the lines you see in the top drawing are shown in their true lengths. This includes the length of line AB. We obtained this top view by projecting perpendiculars from the object to a horizontal plane and then by connecting the points on that plane between each perpendicular. Lines a, b, c and AB are true line lengths.

Now what about the front view? Lines e and f are true lengths, but line AB is foreshortened. The reason for this is that line AB in the front view is *not* parallel to the vertical plane on which it is projected.

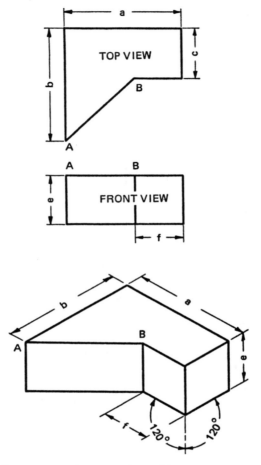

Fig. 6-20. The two upper views are orthographic; the lowest is isometric. Line AB does *not* show its true dimensions in either the orthographic front view or in the isometric drawing. The true length of line AB appears only in the top view.

Examine the bottom drawing in this group. All the lines in this drawing are parallel to one of the three axes and so are true line lengths. But what about line AB? It is not parallel and so this line, line AB, in the isometric drawing does not appear in its true length.

Still considering Fig. 6-20, how many drawings of this object are required? For example, is an auxiliary view needed? The item does have a slant edge, but the top view shows its true size. A side view, not shown in Fig. 6-20, would not contribute any useful information. Actually, all required dimensions could be indicated in the two orthographic views. In this example, the isometric drawing is of no value except to indicate what the object looks like, something you possibly might have been able to visualize with the two orthographic views.

Axis Adjustment

The isometric drawing in Fig. 6-20 may seem as though we have just two axes and indeed, in that illustration, we show two axes displaced by 120°. But what about the third axis, the vertical axis? Figure 6-21 is a redrawing of this part of the isometric and as you can see, the vertical axis is displaced only 60°, not 120°, from the width axis. It has a similar angular distance from the depth axis. However, we can consider the vertical axis as the continuation of a line extending from below the horizontal line. This line, shown as OA in Fig. 6-21, can be imagined as pivoted at point O. It has an angular displacement from the width and depth axes of 120°. We have rotated this line or, if you prefer, just extended it above the horizontal axis. Naturally, we could also extend either the width or the depth axes.

ISOMETRIC PROJECTION

Isometric projection is a form in which we see all three surfaces of the object in just a single view. In Fig. 6-22 we have a cube suspended in space. In front of it we have a plane of projection. The plane of projection, of course, is an imaginary sheet of paper or glass. From each of the visible corners of the suspended cube, we have parallel lines extending to and making tiny dots on our plane of projection. By connecting these dots with lines we get the isometric projection

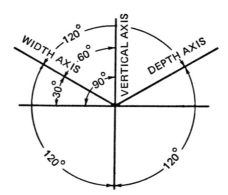

Fig. 6-21. We can consider the vertical axis as a direct continuation of the perpendicular line A_0 from below the horizontal axis.

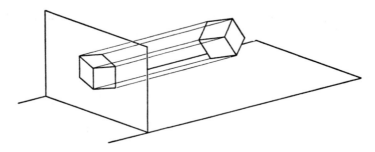

Fig. 6-22. Isometric projection of a cube.

of the cube. Note that we have just a single plane of projection, not two or three.

The object, the cube in this case, is inclined so all of its surfaces make an angle of 35° 16' with the plane of projection. Because of the inclination of the object, the length of each of the edges in the projection on the plane will be somewhat shorter than the actual length of the edge on the object. This reduction is called foreshortening. The amount of foreshortening is equal to 1 divided by the cosine of the angle of inclination (35°16'). The cosine of 35°16' is 0.8165, something you can verify by consulting any table of trigonometric functions. The amount of foreshortening is therefore:

$$1 \div 0.8165 = 1.2247.$$

The significance of all this, without getting too involved in arithmetic, is that if a cube has a side that is precisely 1 inch long, then the projected edge will be only 0.8165 inches long. We get this value by dividing the true length of a side of the cube by the amount of foreshortening, 1.2247, so:

$$1 \text{ inch} \div 1.2247 = 0.8165 \text{ inches.}$$

However, you should keep in mind that all of the surfaces of the object have the same angle with the plane of projection. All of the edges in the isometric projection are reduced by exactly the same amount. Remember, we did not suspend the cube in any position out in space, but deliberately positioned it so all surfaces were at an angle of 35°16' with the plane of projection. As a result, we can use one scale for the projection, using the word "isometric" which means single scale.

ISOMETRIC PROJECTION VERSUS ISOMETRIC DRAWING

In an isometric projection the angle between the three axes on the plane of projection is 120°. Actually, then, except for the difference

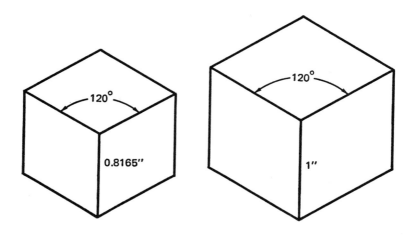

Fig. 6-23. Isometric projection of a cube (left) and isometric drawing (right).

in size, an isometric projection is identical to an isometric drawing. Assume, for example, that the cube we have been working with has a side dimension of 1 inch. Since it is a cube, then its width, length and height are each 1 inch. And those are the dimensions that would exist in the case of an isometric drawing. But if you obtained the cube through isometric projection, each side would be only 0.8165 inch.

In Fig. 6-23 we have a comparison between an isometric projection and an isometric drawing of a cube having a true dimension of 1 inch. The projection is the cube at the left; the drawing is the cube at the right. In the drawing all lines are parallel to the three axes and so are true lengths of 1 inch. And all the lines have the same lengths. In the projection, all lines are 0.8165 inch. The isometric drawing has lines which are about 22 percent longer than those of the isometric projection, but there is an even more important difference. No special scale is required for an isometric drawing. You can lay off an isometric drawing using a regular scale calibrated in inches.

In practical drafting isometric drawings are used more often than isometric projections. In an isometric drawing the foreshortening characteristic of isometric projection doesn't exist, and the dimensions on the drawing are the same as those of the object. However, for a drawing other than full scale, lines can be scaled up or down depending on the scale selected for the drawing. Figure 6-24 supplies a further comparison between an isometric projection and an isometric drawing.

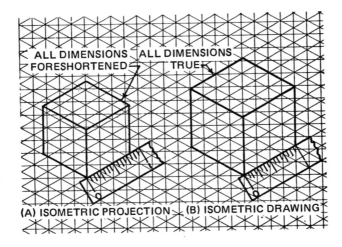

Fig. 6-24. With an isometric projection all dimensions are foreshortened. In an isometric drawing all dimensions are true.

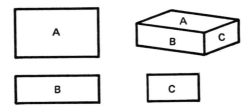

Fig. 6-25. Try drawing three views of these objects.

THINGS TO DO

- Draw three isometric axes and indicate the angular separation of each in degrees.
- Imagine you have a rectangular box measuring 2 x 4 x 6 inches. Draw top, front and right side views. See Fig. 6-24.
- You have now placed a cube, 1 inch on each side, on the rectangular box. The sides of this cube are parallel to the box and the cube is centered on top of the box, on the area measuring 4 x 6 inches. Draw top, front and right side views of this two box assembly.
- Try to draw three views of the object shown in Fig. 6-25.
- Try to draw three views of the object shown in Fig. 6-26.
- Try to draw three views of the object shown in Fig. 6-27. After you have completed your drawings, compare your results with our drawing.

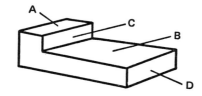

Fig. 6-26. Try drawing three views of this object.

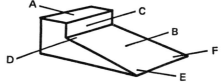

Fig. 6-27. Another practice drawing.

SUMMARY

- In drafting it is customary to examine an object from the top, front and side.
- With central projection, all the projectors come together at a single point.
- A plane of projection is an imaginary flat, transparent surface.
- In parallel projection, the object and its projection have the same size. The lines of projection must be parallel to each other. The object and the plane of projection must also be parallel.
- Parallel projection is called orthographic projection, or orthogonal projection, and sometimes is known as right angle projection. With orthographic projection we work with three views of the object: top, front and side. Sometimes just two views are needed. Occasionally, additional views are used.
- An oblique projection reveals three dimensions: length, height and depth.
- An isometric drawing is a pictorial method that resembles perspective drawing, but with its principal dimensions drawn to scale. In an isometric drawing the dimensions of length, width and height are each represented by an axis. The three coordinate axes are drawn 120° apart. These three axes represent true lengths. An advantage of an isometric drawing is that it enables us to see what the object looks like. We can see all three surfaces of the object in just a single view.
- In an isometric drawing the foreshortening characteristic of isometric projection doesn't exist. The dimensions on the drawing are the same as those of the object.

Chapter 7
Orthographic Projection

The preceding chapter was simply an overview of different kinds of projection. We can now examine one of these, orthographic projection, in much more detail.

There are various ways of drawing orthographic projections but in each the basic idea is the same — to show three views of the same object in their true dimensions. Sometimes, if the object is a simple one, two views will do.

BASIC TECHNIQUES

Just about the simplest object you can draw in orthographic is a cube. All its faces have the same dimensions and so the three views, top, front and side, are identical. Figure 7-1 shows a cube. In this illustration we are looking at the cube through a transparent glass enclosure. We have vertical projectors coming from the cube with these projectors

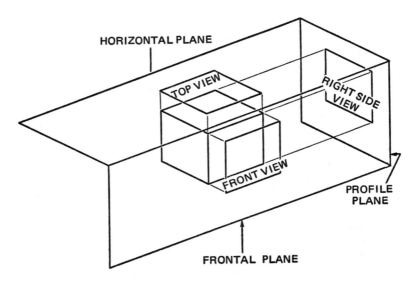

Fig. 7-1. Projection of three views of a cube onto transparent surfaces.

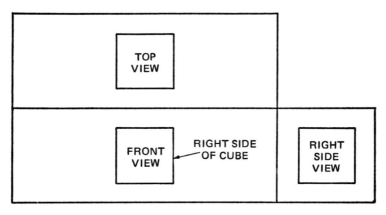

Fig. 7-2. The three views of the cube are now projected onto a single plane.

intersecting the glass. By drawing suitable connecting lines on each section of glass, we obtain the front, top and side views of the cube. In this example the projections, the three views, will have the same size as the object. However, in a drawing we can make these views proportionately smaller or larger.

We can simplify the glass box concept by using the arrangement shown in Fig. 7-2. What we have here are the panes of glass. Instead of showing them in three dimensions, we have them in a single plane, so that all three views are on a single, flat surface. The trouble with Fig. 7-2, though, is that it is too simplified. It does not show how the various views were obtained.

Each pane of glass in Fig. 7-1 is called a plane. We have three of them and we can identify them by name as the *horizontal plane,* the *frontal plane* and the *profile plane.* We can get three views of the object inside the glass box by looking through these planes. We will get a top view if we look directly down from the top, a front view by looking directly through the front, and a side view by looking through the side. Instead of looking, though, we can imagine that the different views of the box are projected onto the planes. These projections will give us our three views: the front view, the top and side views. Sometimes the side view is called the profile view. The horizontal projection is sometimes called the plan view.

We start an orthographic projection by drawing two lines at right angles to each other. These lines can be of random length and are identified as line AB and line CD in Fig. 7-3. The letter O is the intersection point of the lines. We have, in effect, divided the work sheet into four parts, each of which could be called a quadrant and numbered. The numbers are for reference only.

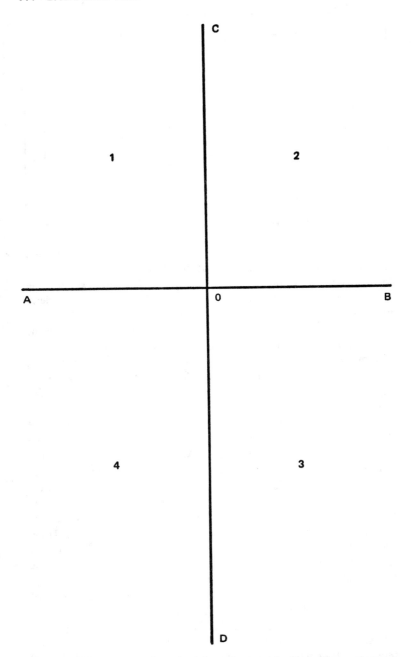

Fig. 7-3. Lines AB and CD are at right angles to each other. The numbers are for identifying the quadrants and do not appear on the work sheet.

And now, as a start, you will need to select one view of the object. Since we are working with a cube all its faces are identical, we can select one face arbitrarily and call it the top view. A top view is one in which we look directly down on an object. Naturally, we will need to know the dimensions of this top view. We can get the dimensions by using a pair of dividers so that the needle points of each leg of the divider rest at the final point of an edge of the cube. Without changing the divider in any way, place the divider needle points on a scale. From the scale you can read the actual dimension.

Note that we do not use a ruler for measuring the object. A ruler is not suitable for our purpose. Furthermore, the use of a pair of dividers and a scale is much more accurate. We can now use the dividers for transferring the measurements of the cube to the work sheet.

In quadrant 1 in Fig. 7-4 draw a pair of fine lines at right angles to each other using your T-square and a triangle. Using the dividers, mark the dimensions of the cube on each of these lines, with one needle point of the dividers resting on the intersection point of the two lines. After doing so, strengthen these lines with your pencil.

Still using your T-square, draw a line that is parallel to reference line AB. This new line can be of random length. It starts at the top end of the vertical construction line of the cube's face. In Fig. 7-5 it is shown as line GH. Finally, complete the view by drawing vertical line JK. Make these light construction lines. After you have finished, use a pencil to emphasize the outline of the face of the cube. We have labeled this face as "top view," although in an actual working drawing these identifying words would not be included.

In Fig. 7-5 the top view has the same dimensions of the object. If the object measures 4 inches on each side but you want the drawing to be 1 inch on each side, you would then work with a 4 to 1 reduction. This means that each 1 inch on the drawing would correspond to 4 inches on the object.

Now that we have the top view we can move ahead and get the bottom and right side views. As a start in this direction, we can extend the existing projection lines of the top view (Fig. 7-6). Note that we have two pairs of such projection lines; one pair is parallel to line AB and the other pair is parallel to line CD. One pair of these projection lines extends down into quadrant 3; the other pair intersect line CD but do not go any further at present.

Figure 7-7 shows our next step with respect to the pairs of projection lines. After intersecting line CD the horizontal projectors make a 45° angle. These projectors continue until they intersect line AB and from that point on are parallel to line CD. Below line AB you

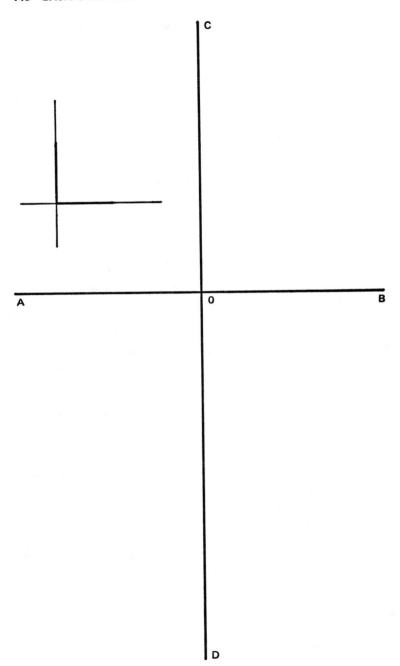

Fig. 7-4. First steps in obtaining the top view of the cube.

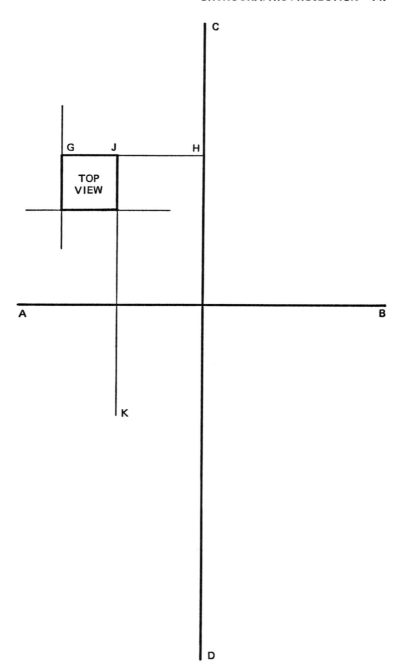

Fig. 7-5. Projection of the top view of the cube.

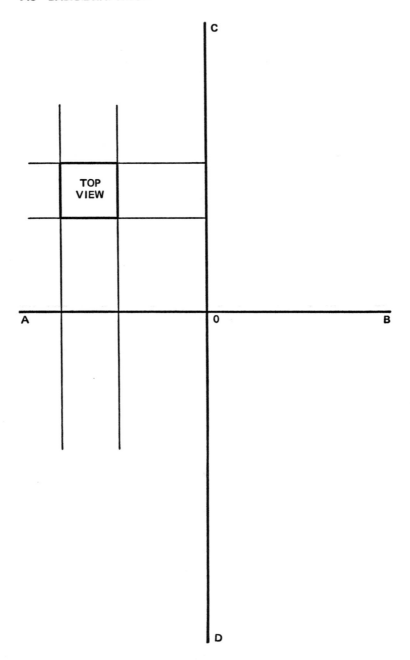

Fig. 7-6. The dimensions of one face of the object are transferred to the work sheet by using a pair of dividers and a scale.

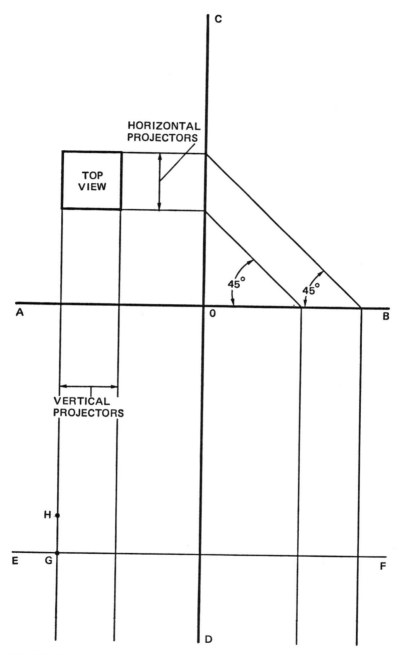

Fig. 7-7. The horizontal projectors from the top view make a 45° turn when they reach quadrant 2.

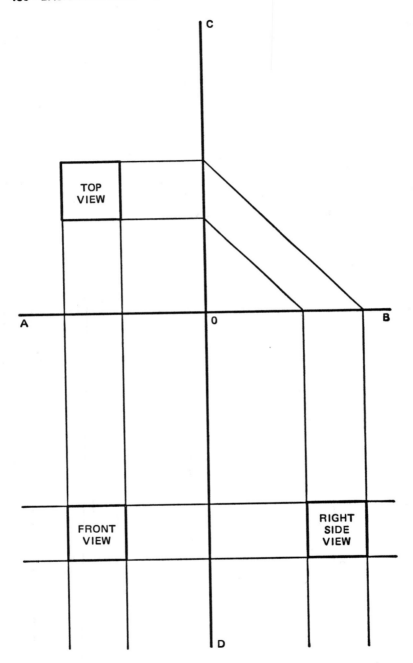

Fig. 7-8. Completed orthographic projection.

can see that the two pairs of projection lines continue straight down and are parallel to each other. All of the projection lines intersect horizontal line EF near the bottom of the work sheet.

At this time, put your dividers at point G. With one of the needle points of the dividers on point G, let the other point of the dividers rest on the line above G. This will give you a needed dimension of the bottom view. Mark this as point H. Draw a horizontal line from this point. When you are finished, you will have the outlines of the front and right side views as shown in Fig. 7-8. Go over these views in pencil until they stand out from the projection lines.

What have we accomplished in Fig. 7-8? We have three views of the object: a top, front and right-side view. We do not see the object in three dimensions since not one of the views gives an impression of depth. There is no perspective. While the views are correct from a drafting point of view, they may seem unnatural for we are accustomed to looking at things in perspective.

PERPENDICULAR POINT PROJECTION

Another method of showing orthographic projection called perpendicular point projection is illustrated in Fig. 7-9. Start with a pair of axes, lines AB and CD, forming right angles to each other at point O.

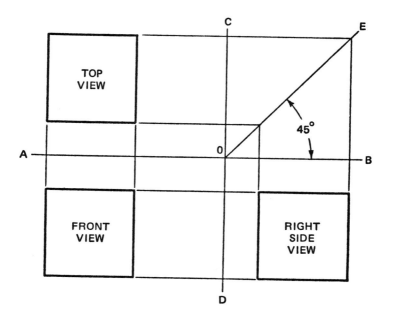

Fig. 7-9. Perpendicular point projection.

From point O draw a line, line OE, which makes a 45° angle with horizontal line AB. Line OE is known as a miter line. When this is done, you will be ready to project the top view.

Draw a pair of parallel lines from the top view so that both of these lines intersect line OE. At the points of intersection, draw a pair of vertical lines for any convenient distance. Go back to the top view and draw a pair of parallel, vertical lines which are actually continuations of the two sides of the top view. All you need do now is to draw a pair of horizontal parallel lines to complete the front and right side views. And, since each face of the cube is a square, the distance from the top to the bottom of the front and right side views will be the same as the distance from the top to the bottom of the top view.

RADIAL POINT PROJECTION

There is still another, easier way of producing the three views of a typical orthographic projection called radial point projection. Start, as usual, with a pair of intersecting vertical lines, shown as AB and CD in Fig. 7-10. Draw a pair of parallel lines extending horizontally from the top view. Set your compass so its point rests at O and adjust the compass until it is exactly at point E. Point E is the intersection of the horizontal projector from the top view with line

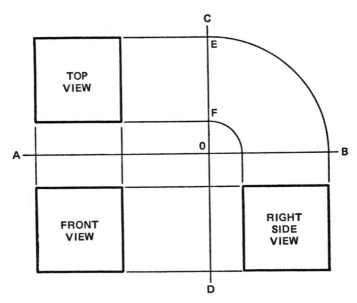

Fig. 7-10. Obtaining orthographic views by radial point projection.

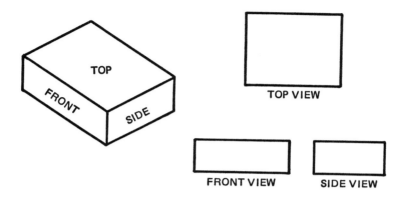

Fig. 7-11. Perspective view of a rectangular box is shown at the left. Orthographic projection of this box is at the right.

CD. Now swing an arc until it intersects line AB. Reduce the opening of your compass. With the point of the compass still resting on intersection O, draw an arc from point F until it intersects line AB.

Orthographic Projection Of A Rectangular Box

The preceding series of drawings were all of a cube, the simplest geometric shape we could start with. But if we can draw a cube we can also draw a rectangular box. The only difference is that some of the dimensions have been changed. Instead of having sides which all have the same measurements, as in the cube, the length, width and depth can be different.

Figure 7-11 shows the orthographic projection of a rectangular box. Projection lines and the miter line have been omitted. Note that the three views are quite similar to those of the cube. Sometimes the locations of the front and side views are interchanged.

Line Weights In Radial Point Projection

When making orthographic projections, use the alphabet of lines described earlier. The two center lines, vertical and horizontal, marked as RL in Fig. 7-12 should consist of long dashes followed by two short dashes although light solid lines are also used. The projection lines from the top view are made of short, rather light dashes. All visible views of the object are solid lines. Note also in

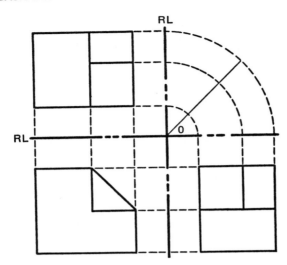

Fig. 7-12. Line weights in radial point projection.

Fig. 7-12 that there are a larger number of projection lines. We have three going across and there are three coming down from the top view. Figure 7-13 shows the same usage of lines with the perpendicular point projection method.

SCALING AN OBJECT

Sometimes an object to be drawn in orthographic projection will be quite large. You may be supplied with the necessary dimensions or

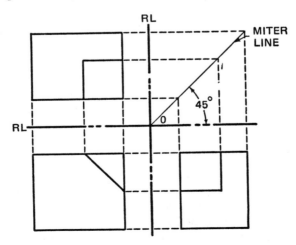

Fig. 7-13. Line weights in perpendicular point projection.

you may need to scale the object yourself. In either case you will need to decide on the reduction scale. The amount of reduction is equal to the actual size divided by the proposed drawing size.

As an example, assume an object has one side that measures 1 foot, 4 inches (1'4") and that you expect to make your drawing of this dimension as 1/2 inch (1/2"). The amount of reduction will be 1 foot 4 inches divided by 1/2 inch. Since the actual size is in two units, feet and inches, we must first convert that size until it is in a single unit, such as inches. One foot is equal to 12 inches. And 1 foot 4 inches is equal to 16 inches. If we divide 16 inches by 1/2 inch we will get 32. Our reduction scale, then, is 32 to 1. This information may need to be put somewhere on the drawing, or possibly in an information block.

The significance of a reduction scale of 32 to 1 (sometimes written as 32:1) is that a dimension of 32 inches on the object is equivalent to 1 inch on the drawing. If we divide both of these numbers by 2 you will see that 16 inches on the drawing is equivalent to 1/2 inch on the drawing. The reduction scale, though, is still 32 to 1.

Orthographic views in preceding illustrations are true views. They represent the actual or true dimensions of the object, drawn the same size, enlarged in scale or reduced in scale. Once dimensions are put on such a drawing you are on your way to supplying the needed manufacturing information.

Orthographic View Of A Cylinder

Figure 7-14 shows a perspective drawing of a cylinder and the three orthographic views. The miter line and projector lines have

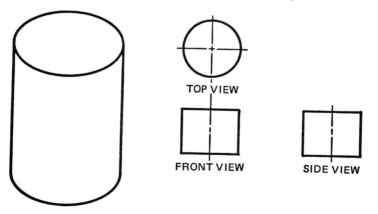

Fig. 7-14. Orthographic projection of a cylinder (right) and the perspective drawing at the left.

been omitted. Note that since the top view is a circle we have used centering lines. The centering lines can be projected over into the front and side views. Both front and side views look rectangular, but it is the top view that supplies us with a clue as to the real shape of the object.

Practice Exercise

A cube is 4 feet on each side. Draw three views of this cube using a scale of 1 inch = 2 feet.

A rectangular box is 3 x 2 x 2 inches. Draw three views of this box in pencil. After the drawing is completed, go over the visible outline in ink. Erase all construction lines.

PROJECTING MORE COMPLEX OBJECTS

A cube is quite a simple object since all its linear dimensions are identical. If you know the length of one edge, you know the lengths of all the other edges. A variation of the cube is the rectangular box, but the basic idea of drawing such a box in an orthographic projection is just the same as that for a cube. Objects that look more complicated may just be combinations of cubes or rectangular boxes. Figure 7-15 is a typical example. What we really have here is a pair of rectangular boxes with the smaller one sitting on the larger. If you are asked to draw such an object, you will either be supplied with the dimensions

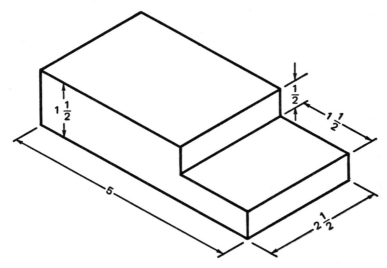

Fig. 7-15. A more complex object may consist of some conbination of simple boxes as shown in this perspective drawing.

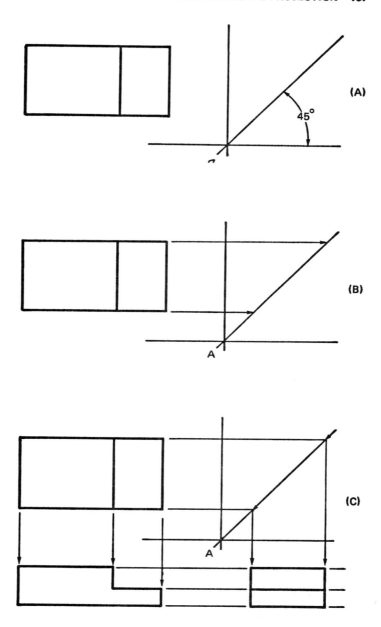

Fig. 7-16. (A). Orthographic projection of the object shown in Fig. 7-15. (B). We have projected horizontal edges of the top view over to the miter line. (C). Projectors are continued down from the miter line to produce the side view.

as shown on the drawing or else you will be handed the object and asked to measure it.

Figure 7-16A shows that we start with a top view of the object illustrated in Fig. 7-15. The vertical line near the center in this view is what you would see looking directly down on the two boxes. Next to the top view we have drawn our vertical and horizontal axes and also a 45° miter line. Figure 7-16B shows how we have projected the horizontal edges of the top view over to the miter line, while Fig. 7-16C shows how projectors are continued down from the miter line to help us produce the side view. Note also how the front view was obtained.

In Fig. 7-16C we used arrowheads on the ends of the projectors, but this was simply to show the direction of movement of the projectors. Normally when making an orthographic projection you would not use arrowheads.

Circular shapes, such as a cylinder, and box shapes like rectangular box, are often combined in one form or another. If you will examine the cylindrical portion first in Fig. 7-17, you will see that its projection is exactly the same as that shown previously in Fig. 7-14. Now examine the rectangular box and you will see that its projection is similar to that in Fig. 7-16. And so an object that at first glance may seem complex may really be nothing more than a combination of simple structures.

The solid shown in Fig. 7-18 has a small cylinder resting on a rectangular solid, one of whose edges has been cut. Now imagine you are looking directly down on this solid. The cylinder would

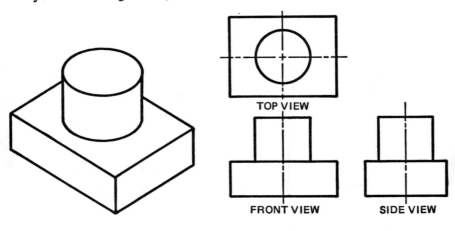

TOP VIEW

FRONT VIEW

SIDE VIEW

Fig. 7-17. Combination of cylinder and rectangular box in orthographic projection.

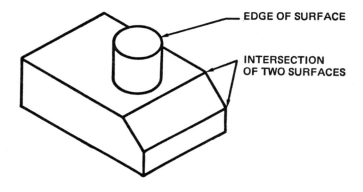

EDGE OF SURFACE

INTERSECTION
OF TWO SURFACES

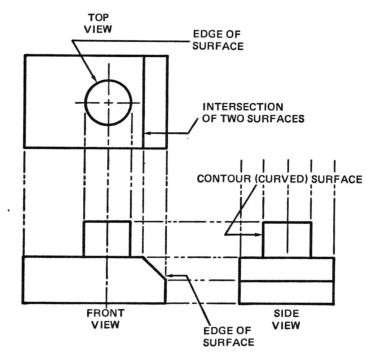

TOP
VIEW

EDGE OF
SURFACE

INTERSECTION
OF TWO SURFACES

CONTOUR (CURVED) SURFACE

FRONT
VIEW

SIDE
VIEW

EDGE OF
SURFACE

Fig. 7-18. Orthographic projection of a more complex solid.

look like a circle and the remainder would seem like a pair of rect-angles. We really have three surfaces and have drawn them as such in the top view. The junction or the intersection of the two rect-angular surfaces is represented by a straight line.

The front view shows the slant edge. The cylinder in this view looks like a rectangle. Even though the top of the cylinder is a circle, in this view, the front view, we are looking at the object "head on" and will see the top edge of the cylinder as a straight line. Similarly, the side view, at the right, also shows the cylinder as a rectangle.

Mental Practice

One of the biggest problems in orthographic projection is to look at three views and then to visualize the object as it really is. It does take some practice. A good technique is to make freehand ortho-graphic sketches before starting an actual drawing. Start with simple objects like a box or any other simple, uncomplicated solid. Make a freehand sketch and then an orthographic projection.

Still another problem is to determine which view to use. In the drawing of Fig. 7-18 we called the part with the cylinder the top, but this was purely an arbitrary choice. Turn the object over and what was the bottom now becomes the top. However, the bottom surface is simply a rectangle, and showing that does not reveal the presence of the cylinder. Select views which either supply the maxi-mum amount of information or the particular information you want.

Simplifying The Problem

While the object in Fig. 7-18 isn't as simple as a cube, actually how complex is it? It's not very complex if you take it apart. It has a cylinder, but a cylinder by itself is quite easy to draw in orthographic projection. It has a rectangular solid and that is also easy to draw. The only variation in the rectangular solid is that it has one slant surface. Whenever you draw an object in orthographic projection, don't look on it as a single complex unit, but rather as an assembly of cylinders, cubes, rectangular solids and so on. If you dissect the object this way, you will find it much easier to draw. You will also find it much easier to visualize from an orthogonal projection.

Projection Of A Curved Surface

The shape of a contour or curved surface can appear as a straight line, as indicated in Fig. 7-18. The curved surface, however, can represent either a solid or can be used to indicate the cutaway section of a solid as shown in Fig. 7-19. If you will examine the perspective

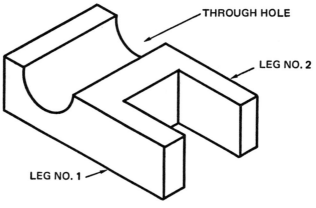

THROUGH HOLE

LEG NO. 2

LEG NO. 1

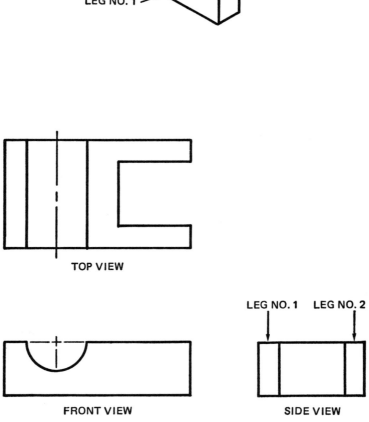

TOP VIEW

FRONT VIEW

LEG NO. 1   LEG NO. 2

SIDE VIEW

Fig. 7-19. Projection of an object containing through hole and cutaway section.

view of the solid you will see that the curve, semicircular in shape, is similar to the cylinder drawn in Fig. 7-17. However, in Fig. 7-19 the curved section has its true contour or curve revealed in the front view, but is shown as straight lines in the top and side views. As you can see, it is very difficult to visualize the object from just one view. With some practice you should be able to do so for simple objects.

One clue to the shape of the semicircular cutaway in Fig. 7-19 is the center line shown in the top and front views. Note also that the semicircular cutaway is a "through hole." It is cut through completely from one side to the other. Also note that the side view supplies no clue that the object has a hole cut through it.

Hidden Lines

Technically, the projection of Fig. 7-19 is incorrect. It is incorrect, however, only in the sense that the projections could be made to yield more information that might be required by the machinist who will produce the object from your drawing. In Fig. 7-19 the object has two legs, marked as leg No. 1 and leg No. 2. You can see both of these legs from the top view, but when you examine the front view, leg No. 1 completely obscures leg No. 2. This is correct. If you were to hold the object directly up against your eyes so that you looked at it "head on," you wouldn't be able to see leg No. 2. Nevertheless, it exists, and so we should show it in our projections. We can do this by using hidden lines, and, as indicated previously by the alphabet of lines, this will consist of a series of short dashes.

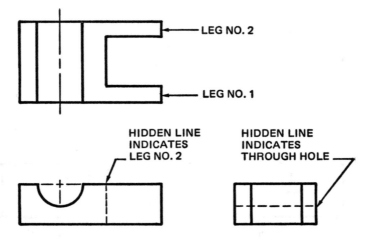

Fig. 7-20. This is the same object as shown in Fig. 7-19 but dashed lines are now used to indicate hidden views.

Examine the front view of Fig. 7-20 and you will see that we have now included a hidden line to indicate the presence of leg No. 2. Now go back to Fig. 7-19 and consider the side view. The side view completely blocks any chance we might have to see the semicircular "hole." Still, the semicircular cutout is there. It does exist and we should know about it. And so we indicate it by a hidden line in our new drawing of the side view. Although the cutout is semicircular, from the side view it looks like a straight line. That is just the way we draw it.

Through Holes

A through hole, as its name implies, is a hole that goes completely through an object. Such holes can be represented by either solid lines or hidden view dashed lines, depending on the view. At the top in Fig. 7-21 is the pictorial drawing of an object and directly below it is the top view. The flat portion of the object has two holes drilled, and these are shown at both ends in the top view. The object in the center that looks like a little hut has two through holes, at right angles to each other. Since we can see the center hole as we

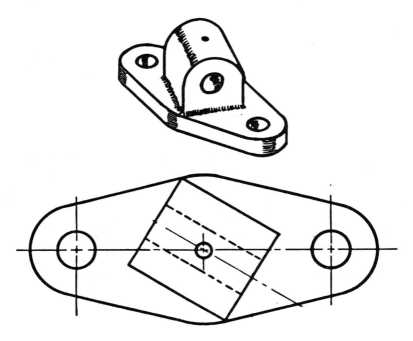

Fig. 7-21. Through holes can be indicated by solid or dashed lines, depending on whether the hole can be seen in the selected view or not.

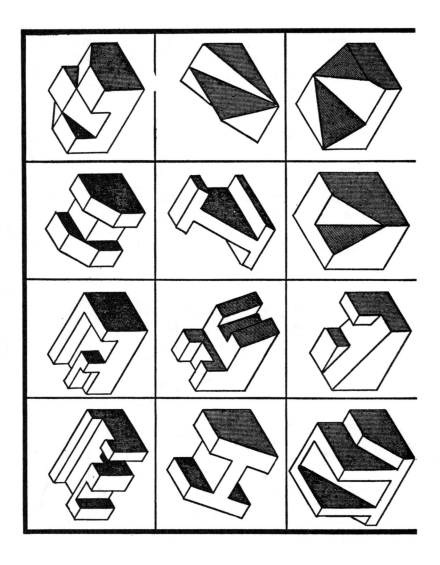

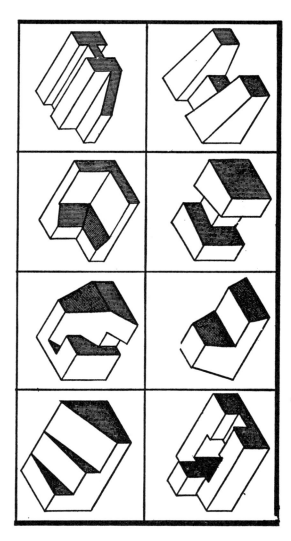

Fig. 7-22.   Geometric shapes for practice drawing.

look down on the object, it is represented by a solid circle. In the top view we cannot see the hole drilled lengthwise through the "hut," and so it is represented by dashed lines.

Drawing Size

A drawing can be smaller than the object it represents or it can be larger. And, of course, a drawing can be made on a 1 to 1 basis, a situation in which the object and the drawing have the same scale. In Fig. 7-21 the drawing is larger than the object.

There is one other thing you should observe in Fig. 7-21. The top view is a visualization of what the object looks like when turned into the position shown in that view. The long axis of the object is made parallel to the upper and lower edges of the drawing paper.

Practice Sketching

A draftsman may sometimes be required to supply projections of an object but without having that object on his table. The object may not be available or it may be fastened to some other object whose size and/or weight make it impossible to bring into the drafting room. It is often helpful for a draftsman to be able to make a free-hand pictorial sketch as a "hands on" substitute.

While the object to be drawn may look complicated, it often may consist of a combination of rectangles, cubes, spheres, other geometric solids or portions of such solids. If you visualize the object in this way, drawing it will become much simpler. As a help in this direction, Fig. 7-22 shows a number of geometric shapes. Select what seems to you to be the easiest to draw and make a pictorial sketch. Then see if you can make a top view. After you complete the top view, see if you can project other views.

SECTIONS

The purpose of an orthographic projection is to supply the maximum amount of information in the least amount of drawing. There is no point in selecting a view just because it is convenient to do so. Thus, in Fig. 7-22 it isn't enough to select views arbitrarily. Some prior thinking is required.

With any object we are interested in two basic bits of information — what it looks like on the outside and inside. The problem is simplified with objects that are solid, but quite often you will also have objects that have holes, or cuts, which require some kind of interior machining.

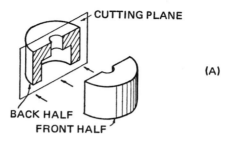

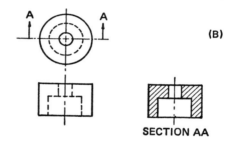

Fig. 7-23. Use of a cutting plane to reveal interior details of an object. (A). A cylinder with a hole cut through its center. (B). Top view of the object.

A symmetrical object is one that can be cut into two or more identical sections. If, for example, an object can be divided into two equal halves, we can take advantage of this fact to supply useful information about the interior. In Fig. 7-23A we have a cylinder with a hole cut through its center. Now imagine we are able to machine our way through this object with a cutting plane. When we finish, the object will be in two parts. Since both halves are symmetrical, we can discard one half and keep the other half for the purposes of our drawing.

Figure 7-23B shows the top view of the object. We represent the cutting plane by a horizontal dashed line going through the center of the object, with this line connecting to a pair of vertical lines marked **AA**. The top view consists of three circles. The outermost circle is a solid line and that is as it should be. Looking down directly on the object shows us the outermost edge. We also see the center hole and that, too, is shown by a solid line. We cannot, however, see the larger circular cutout for it is hidden in the top view. We represent it as a dashed line.

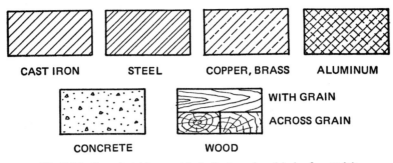

CAST IRON   STEEL   COPPER, BRASS   ALUMINUM

WITH GRAIN

ACROSS GRAIN

CONCRETE   WOOD

Fig. 7-24. Cross hatching used to indicate various kinds of materials.

Now examine the front view, the lower drawing. In this view both holes, the smaller and the larger, are hidden. Hence, we represent them by dashed lines. But what about a side view? Is it necessary? If you were to draw the side view, you would find that it would be identical with the front view. Instead, we draw a section view using section lining or cross hatching to indicate the material of which the object is made. The lines shown in section AA are medium weight, parallel lines, having an angle to the horizontal of 45°, and drawn 1/8 inch apart. This type of cross hatching indicates cast iron, but it is also in general use for all materials in detail drawings. Figure 7-24 shows a few of the American Standard symbols for different kinds of materials.

Partial Section

Sometimes we may be interested in only a small portion of the interior of an object. In that case even a half section might be unnecessary, and so we can use a cutting plane as shown in Fig. 7-25.

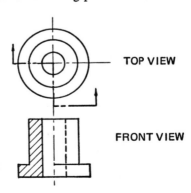

TOP VIEW

FRONT VIEW

Fig. 7-25. In this drawing a cutting plane is used to cut away one quarter of the object.

Try not to visualize the cutting plane as a saw. Instead, think of it as an imaginary shear with which we can slice any portion of an object in any manner. In Fig. 7-25 the cutting plane forms a right angle with itself. Note also that the cutting plane is indicated in the top view. Further, in this plate we have only two views, a top and a front, since the side view would be of no value. The cross hatching indicates that the object is made of cast iron.

Now examine the front view and note the pair of vertical lines extending from top to bottom. The one on the left is a solid line. This is correct since, with the section cut away, one edge of the vertical through hole is exposed. However, the other edge of the through hole is in dashed line form, indicating that it is a hidden edge. Again this is correct for the right hand side if the object has not been cut away.

Offset Section

The cutting plane in Fig. 7-25 forms a right angle, but we can also use it so it forms a succession of straight lines, some of which are offset. If you will examine Fig. 7-26, you will see that the cutting plane starts at one end of the object, forms a right angle, continues straight, forms another right angle and so on.

Examine the top view of this object and you will see that we have a pair of through holes. These holes are offset; they do not use the same horizontal center line. We can use an offset cutting plane to supply more information about these two holes. Note also that the

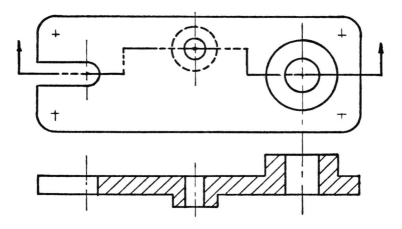

Fig. 7-26. Use of an offset section.

cutting plane lets us use solid instead of hidden lines for these holes. We also know the material of which this object is made.

AUXILIARY VIEWS

You can use one, two or three views for an object. In the case of a solid cube, for example, a single view would be sufficient for all the six sides are identical and no hidden lines are needed. A solid sphere would be another example. For most objects, though, three views are needed. In some cases, not even three views are enough. So we use an extra view known as an auxiliary. Auxiliary views are required in cases when the object has one or more slant surfaces. Projecting such a surface would make the view appear to be foreshortened. To avoid this, we project another view, an auxiliary view, onto a plane surface which is parallel with the slant surface.

If you will go back over the illustrations in this chapter, you will see that in each orthographic projection all three planes of projection are parallel to some surface of an object. This means that each view is projected perpendicular to the plane. The view on the plane is true in size and shape. If you have an object with a slant surface, you should use an auxiliary view if you want that slant surface to have the correct size in the projection. And you can still use top, front and side views for the other non-slant surfaces.

Drawing The Auxiliary View

In Fig. 7-27 we have an object with a slant surface. If you will examine the object, you will see that we have numbered each corner

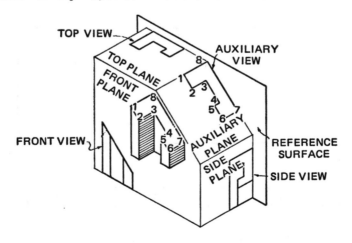

Fig. 7-27. Technique for projecting an auxiliary view.

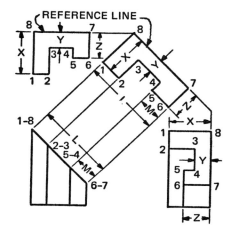

Fig. 7-28. Auxiliary view projection of the object shown in Fig. 7-27.

of the slant surface with numbers ranging from 1 to 8 inclusive. We have surrounded the object with a transparent box and, as you can see, have projected the top view onto the top plane, the front view onto the front plane, and the side view onto the side plane. But we also have an auxiliary plane. To project this view, we draw perpendicular projection lines from each numbered point on the slant surface of the object. These perpendicular lines will pass through the auxiliary plane. We connect the points of intersection. When we do, we will have the true shape and size of the auxiliary view.

Figure 7-28 shows the positioning of the auxiliary view with reference to the other views. To prepare a drawing with such a view, first draw the front and top views, or a front and side view. One of these views must show the edge view of the slant surface. If you will examine Fig. 7-28 you will see that the front view does show the slant edge. We can project the auxiliary view from the front view. Note also that neither the top nor side views show a slant edge. These would not be satisfactory for projecting the auxiliary view.

Now establish a convenient reference line. In the auxiliary view this line will be parallel to the edge of the slant surface. You can then draw perpendiculars from the slant surface to the reference line. Obtain depth measurements using either the top or side views and, measuring from the reference line, transfer them to the corresponding locations on the auxiliary view. You can then join all the points.

There are various letters marked on the drawing in Fig. 7-28, and these appear in the top, right side and auxiliary views. They normally do not appear in a drawing but are used here to help you identify parts of the object in the three views which are the same. The letter Y, for example, refers to identical parts of the object. The numbers that are used correspond to the same numbers in Fig. 7-27.

In Fig. 7-28 the first views that are drawn are the top, right side and bottom views. After these have been obtained, the projection lines can be erased and then the auxiliary view can be projected.

Advantages Of An Auxiliary View

Not only does the auxiliary view show the true size and shape of the slant surface, but you can sometimes use it to help in producing portions of regular views by projecting back from the auxiliary. Another advantage of the auxiliary view is that it often helps in the visualization of the object.

In Fig. 7-29 we have another example of the use of an auxiliary view. The object is an octagonal prism which has been truncated; that is, it has been sliced. As we look down on the object all we will

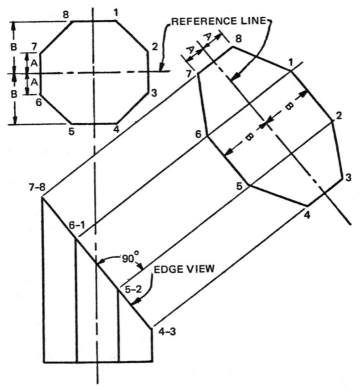

Fig. 7-29. Top, front and auxiliary views of a truncated octagon.

see will be the octagon, an eight-sided figure. The top view does not show the slant and so cannot be used to obtain the auxiliary view. The front view, though, does show the slant. It is from this view that we project the auxiliary. Note that the projectors from the slant are at 90° to it. All points have been numbered so you can check the correspondence of different points in the different views. In a typical drawing, however, these numbers would be omitted. The letters A-A and B-B are used to show you the length of any side (A-A) and the distance between any opposing pair of sides (B-B). The sides of such a geometric figure are often referred to as *flats*.

The example shown in Fig. 7-30 is almost the same as that in Fig. 7-29 except that we now have a truncated cylinder. It isn't as simple a drawing, though, because the auxiliary view doesn't lend itself to instrument drawing. You can produce the top view with a compass, but for the auxiliary view you will need to use a French curve to connect the various points. You will also find it helpful to compare the side views in both Figs. 7-29 and 7-30. They are quite similar if you just consider the edges, but the spacing of the vertical projectors is different.

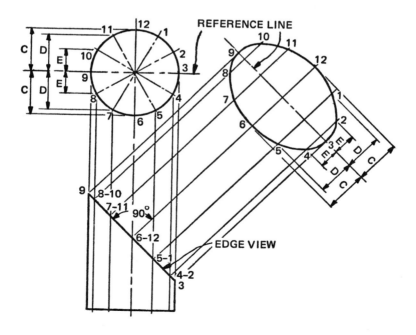

Fig. 7-30. Top, front and auxiliary views of a truncated cylinder.

Fig. 7-31. Sample problems in projection.

Fig. 7-31. Sample problems in projection (continued).

Practice Problems

Figure 7-31 shows a number of objects with some sketches of possible views. As a practice exercise, try to draw the orthographic projections. Also, see if you can supply an auxiliary view, if needed. The views shown in Fig. 7-31 may or may not be correct. If correct, they may have missing outlines or hidden views. Also try drawing a freehand sketch. Whenever you draw any view, try to visualize the view in your mind first. See if you can get a "mental picture" of what the object looks like in each view.

THINGS TO DO

- Make a copy of the object shown in Fig. 7-32. Do not make a tracing. Include all dimensions. Letter the drawing.
- Make a copy of the drawing shown in Fig. 7-33. Do not make a tracing. Include all dimensions. Letter the drawing.
- Two views of an object are shown in Fig. 7-34. The hole that is shown is a through hole. Is the side view correct? Does it have any missing lines? With the help of these two views, try to project a third view.

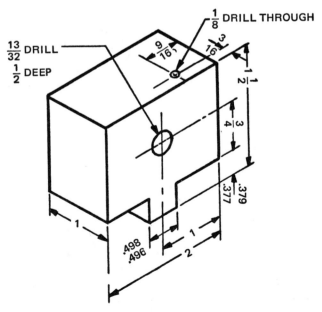

Fig. 7-32. Copy this object. Include dimensions and letter the drawing.

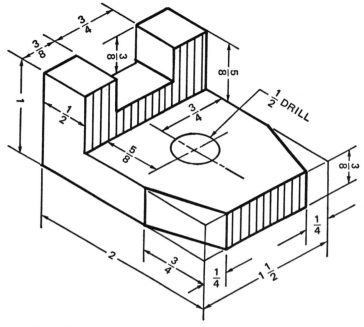

Fig. 7-33. Another drawing for you to do. Don't make a tracing.

- Two views of an object, and the object, are shown in Fig. 7-35. Using this information, try to draw the third view. Draw dimensions on the third view if you think they are needed. Draw an auxiliary view.
- A single view of an object is shown in Fig. 7-36. Draw two additional views, using orthographic projection. Dimension the drawing and assume any dimensions you wish.
- Select one or more objects from those shown in Fig. 7-37 and do a three view orthographic projection. Assume any dimensions you wish.
- Select one or more objects from those shown in Fig. 7-38 and do a three view orthographic projection. Assume any dimensions you wish.

SUMMARY

- The purpose of orthographic projection is to show a number of views, usually three, in true dimensions.
- The three views of an orthographic projection are the top, front and side views.

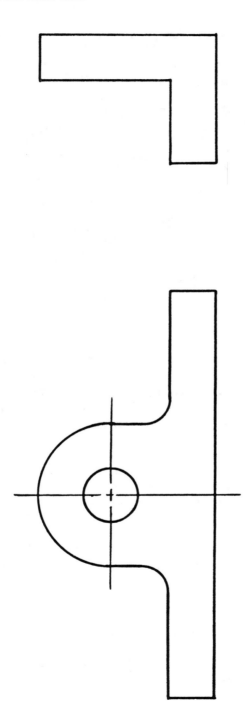

Fig. 7-34. Two views of an object. The hole that is shown is a through hole.

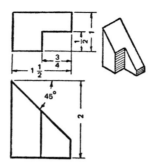

Fig. 7-35. Try to draw a third view and an auxiliary view of these objects.

- Horizontal projectors are used in orthographic projection.
- Objects are not always drawn to scale. A reduction scale is sometimes required.
- Lines of projection do not use arrowheads, but these are sometimes used in rough sketches to indicate their direction.
- It is sometimes difficult, even when supplied with three views in an orthographic projection, to visualize the actual object.
- In orthographic projection, depending on the view, a curved line or circle can appear as a straight line.
- A through hole is a hole that goes completely through an object.
- A section supplies information about the interior of an object. Sections are produced by cutting lines. A section can be half of an object, one quarter, or any other portion, such as a partial section. An offset section is produced when the cutting plane does not follow a straight line, but follows a succession of different paths.
- An auxiliary view is needed to show a true view of the slant surface of an object.

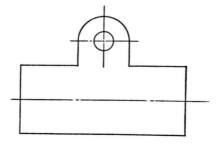

Fig. 7-36. A single view of an object. Draw two additional views using orthographic projection.

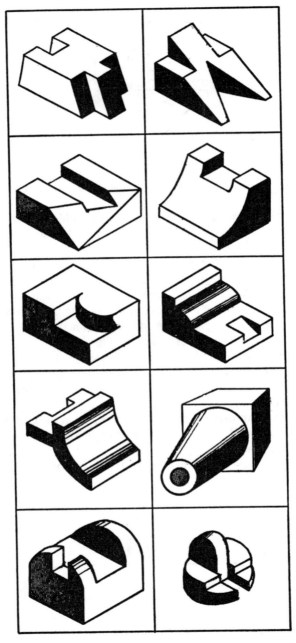

Fig. 7-37. Select one or more objects from these and do a three view orthographic projection.

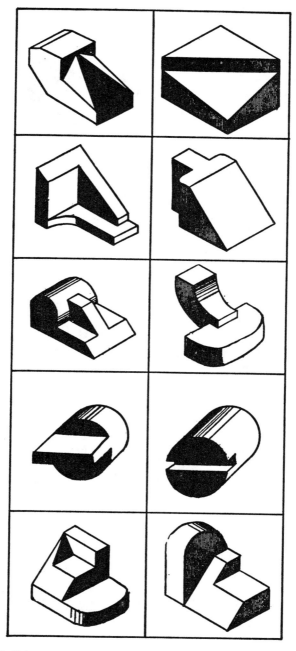

Fig. 7-37. Select one or more objects from these and do a three view orthographic projection (continued).

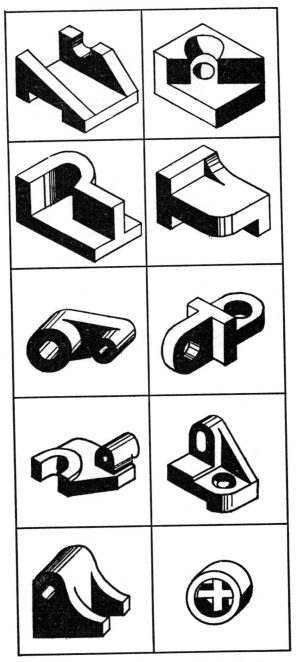

Fig. 7-38. More shapes for you to practice drawing.

Fig. 7-38. More shapes for you to practice drawing (continued).

Chapter 8

Isometric Drawing and Projection

Orthographic projection has its advantages and its disadvantages. It can be made to supply three true views of an object, but often these three views are not enough. We must sometimes call for an auxiliary view. But we may also need additional auxiliary views. Consequently, a drawing might become rather complicated, not only for the draftsman, but even more so for the hobbyist, machinist, contractor or manufacturer who must work from such drawings.

Various techniques have been devised either to supplement orthographic projections or to replace them. One such technique is called isometric drawing.

MAKING AN ISOMETRIC PROJECTION

Making an isometric drawing is comparatively easy. All you need do is to select the three axes representing the three different dimensions, draw these three axes and then construct lines parallel to the axes.

You may also find it necessary to be able to make an isometric projection. One technique is illustrated in Fig. 8-1. The drawing once again is that of a cube and is shown in orthographic projection at the left in Fig. 8-1A. Each edge of the cube is numbered. If you were to draw a straight line from 2 to 4 in the top drawing at the left, you would be creating a face diagonal, which is a diagonal line across the face of the cube. We can also have a body diagonal. This is a line that is drawn through the interior of the cube. We cannot show such a diagonal in any of the orthographic views for these are views of the faces only.

Let us take this cube and rotate it through an angle of 45°. If you will examine the top drawing of Fig. 8-1B you will see that we are still looking down on the cube, but have turned it counterclockwise by 45°. The effect is as though we were looking down on the cube, possibly resting on a table, and that we had just turned the cube by this number of degrees. Except for this angular positioning difference, the top view in Fig. 8-1B is identical with the top view in Fig. 8-1A.

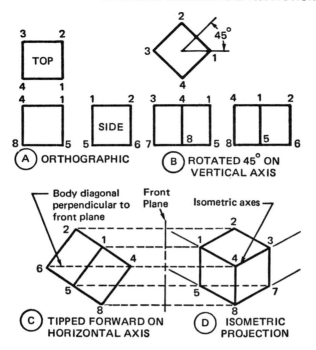

Fig. 8-1. (A). Orthographic projection of a cube. (B). The cube is rotated on its vertical axis by 45°. (C). It is then tipped forward on its horizontal axis. (D). It is projected to produce an isometric projection.

In Fig. 8-1B we have another orthographic projection, but because we have rotated the cube we can now see some sides that were invisible in the first projection. If you will examine the side view in Fig. 8-1B, you will see that we can draw a body diagonal. For example, if we were to draw a line from point 4 to point 6, the line would go directly from one point to the other through the body of the cube. We could also draw a body diagonal from point 2 to point 8.

We are now going to ask you to rest the cube on the point marked No. 8 in Fig. 8-1C. This means the cube will no longer be resting flat on the table but will be tilted at an angle to it. To the right of the cube we are going to ask you to imagine a plane. We are going to look at this plane edge on and so it will simply look like a vertical line. Identified as the front plane, this vertical line will be perpendicular to the bottom edge of your drawing paper. It will form right angles with it.

We have a body diagonal going from point 6 to point 4. This body diagonal must be perpendicular to the line representing our front plane. When we do this, our cube will be tipped the proper angular distance. Now draw a line from point 8 to the front plane and continue it an equal distance on the other side of the front plane. Thus, if you decide to make this dashed line 3 inches from point 8 to the front plane, continue it for another 3 inches. Mark the end of this line with a dot and identify it with the number 8.

At point 8 erect a vertical line. Make this line any distance you wish, but reasonably long. In the tilted cube at the left we have a body diagonal running from point 6 to point 4. Continue this line horizontally until it intersects the vertical line you have drawn from point 8. You will now have a line running from point 8 to point 4 and this will be the vertical axis of the cube at the right (Fig. 8-1D). With point 8 as its apex, draw an angle of 120° as shown in Fig. 8-1D. The arms of this 120° angle are lines 5-8 and 7-8. Do the same at point 4. If you will now draw horizontal projection lines from the remainder of the numbered points in Fig. 8-1C, they will intersect the projections of these two 120° angles. We will be able to establish the corresponding numbered points on the projection in Fig. 8-1D. If you connect these numbered points, using a heavier weight of line, you will get the isometric projection of the cube.

The isometric projection has a number of advantages. Unlike orthographic, it is easy to see immediately what the object looks like. Further, line 4-8, representing the height of the object, is shown in its true length. So are the width, line 5-8, and depth, line 7-8. These three lines are called the isometric axes. Lines parallel to the isometric axes are called isometric lines and also show true measurements.

MAKING AN ISOMETRIC DRAWING

Quite often, a straight line object may seem to have little relationship to the convenient cubes we've been telling you about. A second look will sometimes reveal that the object consists of a series of cubes or rectangularly shaped solids. Figure 8-2 is a representative example.

The drawing at the left is the orthographic projection. Dimensions are marked on this projection in inches. We can eliminate the awkward combination of whole numbers and fractions by converting them to equivalent values in eighths. Thus, 1 inch is equal to 8/8; 1/2 inch = 4/8; 3/4 inch = 6/8; and 1/4 inch = 2/8. The drawing at the right is the isometric. Start with three axes forming angles of 120°. Swing the altitude up so that it represents a perpendicular joining the width and depth axes. Now measure off the length of the width axis. This is shown in the drawing as 20 and of course it means 20/8. Also draw the depth axis and it should be 12/8. The

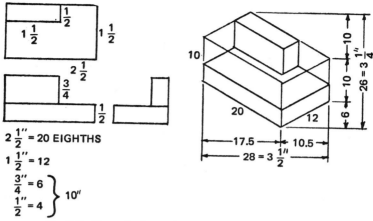

$2\frac{1''}{2}$ = 20 EIGHTHS

$1\frac{1''}{2}$ = 12

$\left.\begin{array}{l}\frac{3''}{4} = 6 \\ \frac{1''}{2} = 4\end{array}\right\}$ 10″

Fig. 8-2. Dimensioning technique in an isometric drawing.

actual lengths of these lines could be 20/8 and 12/8, or you could scale these numbers down or up. Multiply them by two and you have times two scale. Divide by two and you have half scale. Use any convenient multiplier or divider you want, depending on just what size you want for the final drawing.

After you have completed the three axes, you can complete the box. Examine the smaller box shown in the top orthographic projection as being 1/2 x 1 1/2. This smaller box has its own axes, but these are parallel to the axes drawn for the larger box, and they too have an angular separation of 120°.

When the drawing is completed, erase the extension lines that appear above the larger of the two boxes. The isometric drawing will be complete. You can make an isometric drawing using the information supplied by an orthographic projection or else you can take data supplied by the object itself. One of the things Fig. 8-2 does demonstrate is that you do have a choice. You can either draw in orthographic or in isometric. However, you may not always be able to draw the way you prefer, for the choice may be dictated by the rules of a drafting department. However, if you do construction work of your own or you have a hobby that requires plans, then you can use the method you find easier and more appealing. Dimensioning on the orthographic projection is cleaner and neater, but the isometric does supply a more realistic picture.

Figure 8-3 shows a rectangular solid that has had a section removed. Make an orthographic projection of this object and then do an isometric sketch. Also try to make an isometric projection.

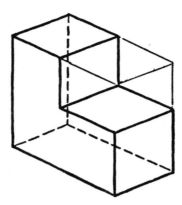

Fig. 8-3. Practice exercise in orthographic projection and isometric projection and drawing.

INVISIBLE LINES

Invisible lines are generally omitted in isometric drawings. Isometric drawings are often used by persons who have had no drafting training and so such lines would tend to be confusing. It all depends, then, on who is going to work with such drawings. If you feel that invisible lines are required and that their function will not be misunderstood, include them. Similarly, center lines are omitted, unless you need them for some purpose, such as dimensioning.

DRAWING RECTANGULAR SOLIDS

The rectangular solid is the basic shape in isometric drawings. This doesn't mean that every object drawn in isometric is a complete rectangular solid. It may consist of a combination of such solids, or may have a cutaway section. However, the basic idea is to construct the outline of a rectangular box to serve as the framework and then, after the isometric drawing is finished, to erase nonessential parts of that framework.

- You will need a pencil, T-square, 30°-60° triangle, scale and drawing paper.
- Draw a line horizontal to the bottom of the paper and at any convenient point along that line put a tiny pencil dot (Fig. 8-4). You can call this zero or your starting point.
- Put your T-square so that its head rests against the left side of your drawing board. Use your triangle to draw a vertical line through the zero point. Make sure this line extends above and below the horizontal line.

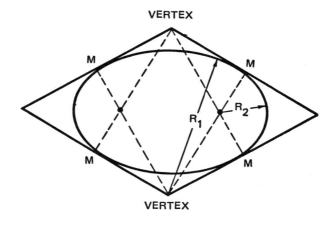

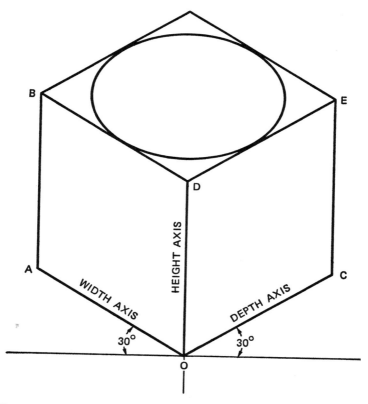

Fig. 8-4. Technique for drawing an isometric circle on the top face of a cube.

- With the help of your T-square and 30°-60° triangle draw a pair of lines (OA and OC) so these lines slant upward from the zero point, making an angle of 30° with the horizontal. The left slant line will be your width axis; the right slant line will be your depth axis. You have already drawn the vertical (height) axis.
- Measure along the three principal axes with a scale, always starting from zero on the drawing. You will be measuring height along the vertical axis, width toward the left of the vertical axis and depth toward the right. Draw these three axes.
- The ends of each of these lines — the width, depth and height lines — are known as terminal points (points A, C and D). And now that you have these terminal points, use your T-square and triangle to erect lines parallel to the main axes and having the same dimensions. When completed, you will have a rectangular isometric box. You can draw additional lines inside this box to show the outline of the actual object. Any such lines parallel to the three main axes will be in true dimensions. Any other lines will not be in accurate dimension.
- For any lines to be drawn inside the isometric box which are not parallel to the main axes, establish coordinate points from projectors from an orthographic projection.

ISOMETRIC CIRCLES

Isometric circles are always an oval or an ellipse. Think about it for a moment and you will see why. We never look at an object "head on" when working in isometric. In orthographic projection we have the surfaces of the object parallel to imaginary planes. But in an isometric projection or drawing we view the object "edge on." There is only one plane and that plane is parallel to the joined edge of the three axes of width, depth and height. This means that any circle on any face of an object viewed in isometric will also be seen at an angle.

To understand the meaning of all this, draw a circle on a sheet of cardboard and then cut out the circle. Hold the circular section of cardboard so that you face it directly. If you will try to keep your eyes on the outer circumference, you will see a full circle. Now slowly tilt the circular cardboard. As you tilt it the circular shape will begin to take on an oval or elliptical shape. After you have made a complete 90° turn, all you will see will be the edge of the cardboard. Since this edge is the circumference of the circle, it will appear as a straight line. When you draw a circle in isometric, it will be equivalent to your circular cardboard being held at an angle to your eyes instead of facing you directly.

Drawing An Isometric Circle

In Fig. 8-4 we have the cube drawn in isometric and an isometric circle on its top face. The upper drawing shows how the isometric circle is constructed. If you will examine this "circle," you will see it has four points identified by the letter M. Each of these is a midpoint of a straight line edge of the cube. Use your 60° triangle to construct perpendiculars to these points. These lines, shown in dashed form, should extend directly into the opposite vertex. There will be two such corners and they will be diagonally opposite. Use the vertex for the needle point of your compass and draw a large arc with a radius R_1 as shown in the upper drawing. To draw this arc set your compass so that it extends from the bottom vertex to M on the opposite side. Draw an arc so that it goes from M at the top left to M at the top right. Move your compass to the opposite vertex and repeat, drawing an arc from M on the bottom left to M on the bottom right.

The dashed lines you constructed will intersect at two points. Select one of these points and adjust your compass so that its opening is equal to the distance from the intersecting point to M. This radius is identified as R_2. Now draw a radius from M at top right to M at bottom right. Repeat this action on the left side and your ellipse should be complete.

You can draw an ellipse on any of the three visible faces of the cube. Figure 8-5 shows what the cube would look like with ellipses on each of the faces.

Figure 8-6 illustrates the technique to use for drawing an ellipse on one of the vertical faces. The method is the same as that just

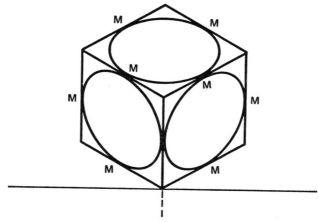

Fig. 8-5. Isometric cube with isometric circles on each face.

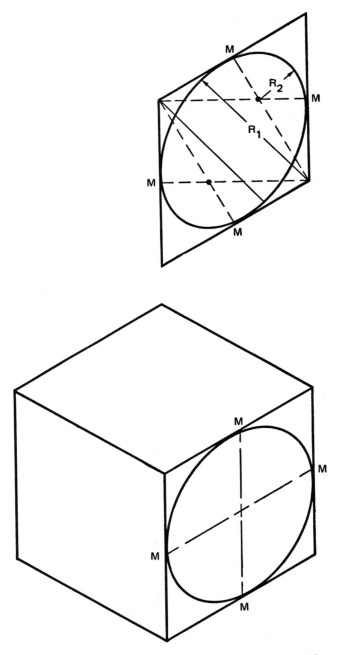

Fig. 8-6. Technique for drawing an isometric circle on either vertical face of a cube.

described. Locate center points M and draw dashed lines to opposite corners. Use a corner where a pair of such lines meet and set your compass equal to the length of the dashed line from M to the corner. Use this length, shown as R_1, to draw a pair of large arcs. The intersection points, shown as R_2, will give you the dimension of the smaller arcs. You will need to draw two large arcs and two smaller ones. When finished you should have an ellipse. If you have an open space between any pair of drawn arcs, simply join them with the help of a French curve.

You can see the relationship between the circle in orthographic and the same circle when drawn in isometric by examining Fig. 8-7. In Fig. 8-7A we have an orthographic view of the face of an object. Dimension D is the true dimension. Note that the orthographic view of the circle uses a pair of center lines.

Figure 8-7B is the first step in the isometric drawing. Measurement D is exactly the same as measurement D in the orthographic view. Figures 8-7C, 8-7D and 8-7E are produced using the technique described in connection with Fig. 8-6. The lines which form perpendiculars with the edges of the object in Fig. 8-7C are identified by

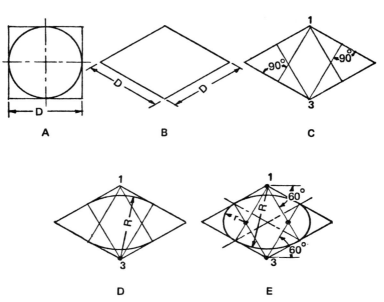

A   B   C

D   E

Fig. 8-7. Method for drawing an isometric circle on any face of an object drawn in isometric. (A). Orthographic view of the face of the object. (B). First step in the isometric drawing. (C). Lines which form perpendiculars with the edges of the object are identified by the 90° dimension. (D). You can see the 60° angles formed. (E). Ellipses on the faces of the cube also have center lines.

the 90° dimension. The fact that these lines may not appear to be perpendicular is an optical illusion. When erecting such lines, you do not need to include the 90° dimension.

Figure 8-7E shows that the ellipses on the faces of the cube also have center lines. You can construct such center lines by finding the midpoint of each edge and then drawing a line to the midpoint of the opposite, facing edge. To see this more clearly, go back to Fig. 8-6 for a moment and examine the lower drawing. The crossed dashed lines from M to M are the center lines.

Smaller Circles In Isometric

Figures 8-6 and 8-7 seem to imply that the circles in isometric must occupy the full surface areas. Actually, an isometric circle can have any dimensions, just as it can in orthographic projection.

You can also draw one isometric circle inside another. Such circles will be concentric to each other.

Drawing A Cylinder In Isometric

Figure 8-8 shows how you can draw a cylinder in isometric. As a start, construct an isometric ellipse for one end. The important points in this isometric ellipse are A, B and C for it is with the help of these points that we are able to draw the large and small arcs composing the ellipse. Draw lines from these points in a perpendicular manner; that is, each of these lines should be parallel to the vertical axis of the cylinder. The vertical axis of the cylinder is an

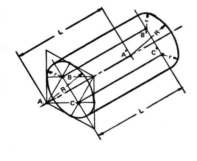
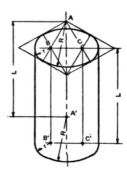

Fig. 8-8. Technique for drawing a cylinder in isometric.

imaginary line running through the full length of the cylinder for the distance L shown in the two drawings. The three lines drawn from points A, B and C should have a length equal to the height of the cylinder. The terminal points of these three lines will be A′, B′ and C′. With the help of these points, draw the large arc and the two smaller arcs to form the ellipse for the bottom of the cylinder. The method to use here is exactly the same that you employed for drawing the ellipse at the top of the cylinder. The only difference is that the bottom ellipse will not be seen in its entirety.

All you will need to do to complete the cylinder will be to draw a pair of lines representing the outer edges of the cylinder. You will, in effect, be connecting the upper ellipse with the lower one.

If you will go back to Fig. 8-7E, you will see that you have three possible positions for the ellipse. This means you should be able to draw isometric cylinders so they appear in these three different positions. Figure 8-8 shows two of these. As a practice exercise, try drawing an isometric cylinder in the third position.

DRAWING AN ISOMETRIC ARC

One way of drawing an isometric arc of any dimension less than a complete circle is to draw a complete isometric circle and erase that portion of it that isn't needed. While this is a simple approach, it also means you must do the maximum amount of work, most of which must be discarded.

Another method which you can use for drawing a quarter circle appears in Fig. 8-9. Draw a pair of lines, AD, and BD of equal length,

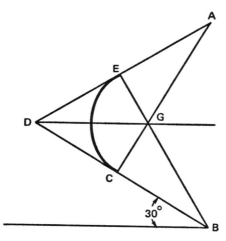

Fig. 8-9. Method for drawing a portion of an ellipse.

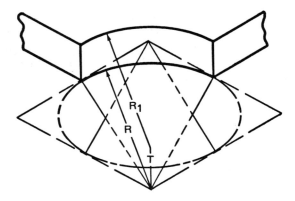

Fig. 8-10. Use of an ellipse to draw a portion of an object in isometric.

terminating in points A and B. An imaginary line drawn from points A to B would be perpendicular to the horizontal line across the bottom. Lines AD and BD can be of any length, depending on the size of arc you want. From the center point of AD (shown as point E), draw a line so it rests on point B. This line should be perpendicular to line AD. Similarly, locate the center point of line DB and erect a perpendicular at point C. It should terminate at point A. These two lines will intersect at point G. Set your compass to the distance EG and draw the required arc.

We still have a problem, though. If you will examine an ellipse, such as the one shown earlier at the top in Fig. 8-4, you can see that the curvature of the ellipse, unlike that of a circle, varies sharply. Which section of arc you will use will require judgment on your part. Figure 8-10, for example, shows the construction of an arc to join the ends of an object drawn in isometric. Actually, any part of the ellipse could have been used and the result would have been a greater or lesser curvature of arc. In any event, since the arc cannot possibly be parallel to the three main axes of the isometric drawing, the arc is not a true dimension.

DRAWING CURVES AND IRREGULAR SHAPES IN ISOMETRIC

While an object such as a rectangular solid may very well have a number of straight lines, or some circles and straight lines such as a cylinder, it is also possible for an object to have an irregular shape. You can draw irregular curves by using the method shown in Fig. 8-11.

As a start we have two orthographic views, the top view (left) and the right side view. Divide the top view into equal segments by drawing

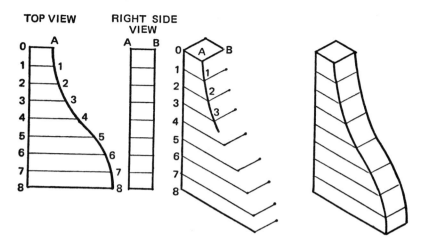

Fig. 8-11. Method of drawing an irregular curve in isometric.

a number of horizontal lines. In the top view there are seven lines drawn horizontally from left to right which divide the object into eight equal areas.

Using horizontal projectors, extend these dividing lines over to the side view. To the right of the side view, draw a perpendicular with this perpendicular divided by extensions of the horizontal projection lines. At point O erect an isometric view of the top of the object. This top is a square since the distance O-A is equal to A-B. The distance O-A in the isometric drawing will be equal to the distance O-A in the orthographic projection.

Starting at point 1 in the isometric, draw a line 1-1 parallel to the line O-A above it. Line 1-1 should have the same length as line 1-1 in the top view of the orthographic projection. Continue by drawing line 2-2 in the isometric, then line 3-3 and so on. When you have finished, connect all the terminal points of these lines with a French curve.

The drawing at the far right of Fig. 8-11 shows how the isometric is completed. Draw a succession of lines parallel to A-B. Each of these lines will have the same lengths as 1-1, 2-2, and so on. Draw the second irregular curve to connect the end points so that the second irregular curve is parallel to the first one. You can transfer line dimensions by marking them off on the edge of a card or slip of paper. This is more accurate and faster than using a compass or ruler. When you are finished, erase all the horizontal construction lines.

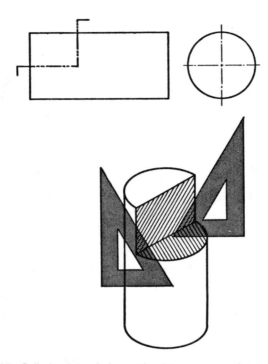

Fig. 8-12. Cylinder drawn in isometric with cutaway section showing hatching. Note that the vertical and horizontal sections are hatched with triangles positioned in opposite directions.

DIAGONAL HATCHING IN ISOMETRIC

To cross hatch a sectional view drawn in isometric, use your 30°–60° triangle. In the upper part of Fig. 8-12 you can see the side and end views of this cutaway cylinder in orthographic projection. The view shown at the top left indicates the presence of a cutting plane. The diagonal hatching should give the appearance of making a 45° angle with the horizontal or vertical axis of the surface. Naturally, if the drawing of the object were only in orthographic, a 45° triangle would be called for, but to give the appearance of 45° in isometric we need to use a 60° slant. If the surface to be hatched is non-isometric, you will need to experiment until you get the effect you want.

OBLIQUE DRAWING

Quite logically, you probably associate drawings with architecture or machinery, but they are also widely used in other applications as well. Drawings used in the furniture industry, for example, are

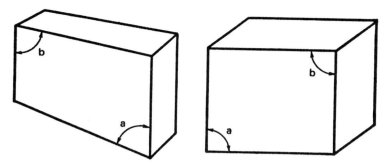

Fig. 8-13. Perspective drawing (left) and oblique drawing (right). This type of oblique drawing is sometimes called cabinet. In the perspective drawing, angle a is not equal to angle b. In the oblique drawing, angles a and b are equal.

known as cabinet drawings. This drawing style, with some changes, is also used by draftsmen but is called *oblique drawing.*

Figure 8-13 shows the difference between a perspective drawing at the left and an oblique drawing at the right. If you will examine the perspective drawing, you will see that the horizontal lines seem to come together as your eyes move from right to left. The portion of the object drawn closest to your eyes appears to be larger. If we could extend the drawing far enough, all the horizontal lines would come together at a point.

Unlike the perspective drawing with the object facing us at some angle, the oblique drawing shows one side, the front in this example, parallel to the paper on which it is drawn. If we were to pass a horizontal plane in front of the object, we would see the front view on that plane with the front view in its true dimensions.

If you will go back to the perspective drawing for a moment, you will see that all the angles are different. Because they are different in the drawing, we cannot measure them. With a rectangular box, such as that shown in Fig. 8-13, we know that each of the angles must be 90°. This is true in the oblique drawing and so we can measure such angles.

Figure 8-14 shows three views of a cabinet drawing — the top and front views and a side view. The length of this box is represented by the letter a. In the top view, the length a is a true measurement. However, in the bottom drawing, length a is reduced by 50 percent. Measurements b and c are true measurements.

Select one side as the face and either draw it full size, if the object is small enough, or in reduced scale (Fig. 8-15). From each of the corners of the front view, draw lines which make an angle of 45° with the horizontal edges of the front view. These lines will represent

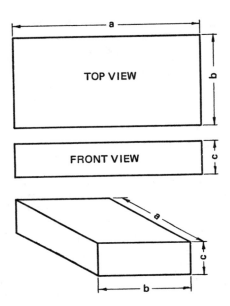

Fig. 8-14. Three views of a cabinet drawing. In the bottom drawing line a is half the size of line a in the top view, although both lines represent the same edge in this box.

the depth of the object. Divide this depth by 2. Thus, if the object has a depth of 2 inches, and you are drawing to scale, the depth lines will be 2 divided by 2 or 1 inch. After drawing the depth lines, draw the rectangle representing the rear view. This rectangle will be parallel to the front view and will have the same dimensions. After completing the rear view the box will be finished. Erase all unnecessary lines.

The receding axes in Fig. 8-15 form an angle of 45° with the horizontal lines of the front view. Other angles can be used, but the

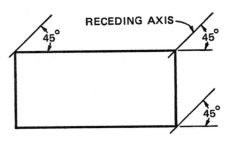

Fig. 8-15. Steps in making a cabinet drawing. Although the angle shown here is 45°, angles of other sizes can also be used. The only requirement is that all angles must have the same dimension.

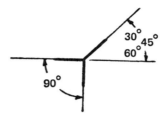

Fig. 8-16. The edges in the front view in an oblique drawing are at 90°. Although 45° is commonly used for the receding axis, you can also use 30°, 45° or 60°.

requirement is that they must all have the same angular opening. In a cabinet drawing the receding axes generally go upward to the right, although you will see some cabinet drawings in which the receding axes are drawn upward to the left. The great advantage of the 45° line is that you can use your 45° triangle to draw it. Other commonly used angles are 30° and 60°. These can also be drawn by using a triangle, in this case a 30°-60° triangle.

An oblique drawing, as shown in Fig. 8-16, always has three axes. Two of these are always perpendicular to each other, and so form a 90° angle. One face of the object is drawn parallel to the plane of projection and is therefore shown in its true shape and size. This raises the question of which face to select. There is a tendency to select the largest surface area, but a better technique is to choose the face having the most detail or which has curved edges or an irregular contour.

Figure 8-17 shows how you can go from an orthographic projection to oblique. Here we have two views. The front view in the orthographic

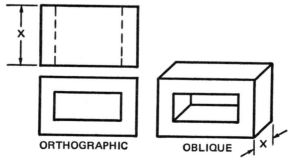

Fig. 8-17. You can make an oblique drawing from information supplied by an orthographic projection. The depth, X, in the oblique drawing should be half the depth X shown in the top view of the orthographic projection.

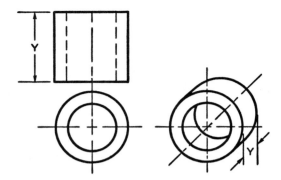

Fig. 8-18. Orthographic projection of an object with a through hole (left) and an oblique drawing at the right.

projection can be the front view of the oblique drawing. Select an appropriate angle for drawing the receding axes, making sure that these edges are half size. If you have an object with a through hole, you can still use the front view of the orthographic projection as the front of the oblique drawing, as indicated in Fig. 8-18.

If you have an orthographic projection and you want to draw an oblique, there is no reason why you cannot transpose orthographic views so that your oblique will be easier to draw. In Fig. 8-19, for example, the view shown at the top would normally be a side view. The bottom view in the orthographic would be the top view, but it lends itself nicely for drawing an oblique. In this example, then, draw the circular parts so they are parallel to the plane of projection. When making the oblique drawing follow these steps: draw the base, stem, hole and the remaining lines.

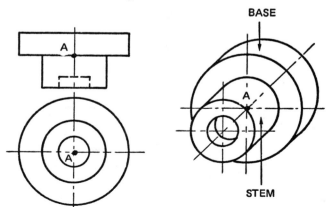

Fig. 8-19. Construction of a cylindrical object in oblique.

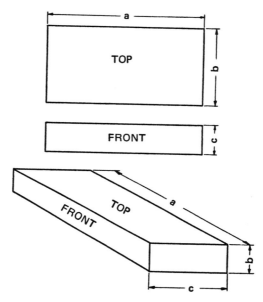

Fig. 8-20. In a cavalier drawing, top, all dimensions, such as a, b and c, are true dimensions.

CAVALIER DRAWINGS

In a type of drawing known as cavalier, all dimensions shown are true. Figure 8-20 shows a rectangular box drawn using the cavalier technique. Dimensions a, b and c in each of the views represent true dimensions. Figure 8-21 shows the basic difference between identical boxes drawn in cavalier and cabinet forms. The box at the left is cavalier; that at the right is oblique. The foreshortening of the depth in the oblique drawing makes the box look more realistic.

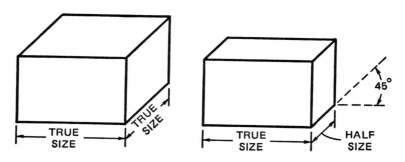

Fig. 8-21. Basic differences between cavalier drawing (left) and oblique (cabinet) drawing (right). These are two drawings of the same object.

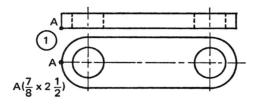

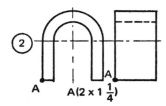

Fig. 8-22 Practice exercises in making an oblique drawing from orthographic projections.

THINGS TO DO

- In Fig. 8-22 you have orthographic projections of two objects. Dimensions are supplied with each or, if you prefer, assume dimensions. Make an oblique drawing using 45° axes.
- Draw a square with edges of 2 inches. Inscribe a circle inside this square so that the circle touches the edges of the square. Then make an oblique drawing.
- Draw an oblique cylinder with a vertical axis of 5 inches and bottom and top circles with a radius of 1 1/2 inches.
- Make a cabinet drawing of a cube whose sides are 2 inches. Make a cavalier drawing of the same object.

SUMMARY

- An isometric projection can be obtained from an orthographic projection by rotating the object through a selected angle, such as 45°.
- With an isometric projection it is easy to see what the object looks like. Lines parallel to the isometric axis are called isometric lines and show true measurements.
- You can make an isometric drawing using the information supplied by an orthographic projection or take the data supplied by the object.
- Invisible lines are generally omitted in isometric drawings.
- The rectangular solid is the basic shape in isometric drawings.

- Isometric circles are always an oval or an ellipse. One isometric circle inside another isometric circle is concentric to it.
- One way of drawing an isometric arc of any dimension less than a complete circle is to draw a complete isometric circle and erase that portion of it that isn't needed.
- A 30°-60° triangle can be used to cross hatch a sectional view drawn in isometric.
- Drawings used in the furniture industry are known as cabinet drawings. This drawing style, with some changes, is called oblique. An oblique drawing always has three axes. Two of these are always perpendicular to each other and so form a 90° angle. One face of the object is drawn parallel to the plane of projection and is therefore shown in its true shape and size.
- In a cavalier drawing all dimensions shown are true.

Chapter 9
Revolutions and Intersections

If you had an object such as the V block illustrated in Fig 9-1 and you wanted to know as much about this block as possible, you would examine it from a number of different angles. You might try looking straight down on the block, and directly at its front and side to plan an orthographic projection. You might decide that this block would also require an auxiliary view. You might also think about doing an oblique drawing. In short, your goal would be to produce a drawing, or drawings, that would convey enough information to enable the block to be manufactured.

However, it is also possible that no matter what technique you used, you would still have the problem of supplying data about the true shape of inclined lines. There are two methods you can choose for obtaining the true shapes of inclined surfaces and true lengths of inclined lines. One of these is the auxiliary view method used earlier in connection with orthographic projection. The other is the method of revolution.

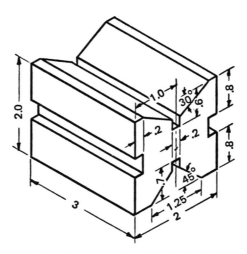

Fig. 9-1. This V block can have a number of slant surfaces which require special drafting treatment.

REVOLUTION OF AN OBJECT

In orthographic projection we have three principal planes of projection. Since it is possible for an object to have a surface which is inclined to one or more of these planes, we can use a special plane of projection which is parallel to the inclined surface. This is the technique we used for getting an auxiliary view. And so, in getting such a view, we catered to the object. The object remained fixed in position and it was our job to maneuver a plane of projection to accommodate the object.

However, there is no reason why we cannot move the object, just as we do when we examine it. That's the method we used when we first started our observations of the V block. There's no reason why we cannot use this technique for drafting. And so what we are going to do now is get a view of the object, showing the true size and shape of an inclined surface by revolving the object until the slant surface is parallel to one of the three available planes of projection. This action on our part is called *revolution.*

Simple Revolution

In orthographic projection we usually have three planes parallel to three selected surfaces of an object. We can imagine a line coming from the object, possibly from its center, and going through the center of the plane. This line will be a perpendicular; that is, it will be perpendicular to the object and also to the plane. We can call this line an axis. If we have an orthographic projection we can revolve the object around an axis which is perpendicular to the top plane, the front plane or the side plane. We can then draw the object in its new position.

In Fig. 9-2A we have the usual orthographic views of a simple, rectangularly shaped object. Now suppose you decide to tilt the front view so that it forms an angle of 30° with the horizontal. What we have done in this case is to revolve the object on an axis perpendicular to the vertical plane. You can see the results of this revolution or movement of the object in Fig. 9-2B. The top and side views have changed as well as the front view. In the orthographic views we projected from the top view to the front and right side views. But when we revolved the block, we also used it to project new top and right side views.

Consider the front view in Fig. 9-2B. All we have done is to tilt the block upward by 30°. The front view is still parallel to its plane of projection and that is why this view remains the same. Compare the front views in Figs. 9-2A and 9-2B and you will see that, aside from the 30° angle, both front views are alike.

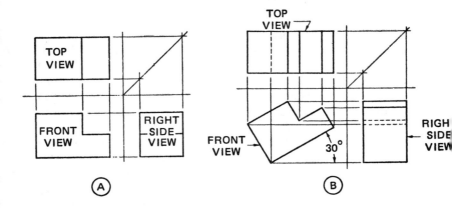

Fig. 9-2. (A). Orthographic view of a rectangularly shaped block. All dimensions in each view are true. With the block tilted 30° parallel to the projection plane of the front view, only the front view dimensions remain true. (B). The top and right side orthographic projections from the tilted front view.

However, if the front view remains unchanged and is still parallel to the plane of projection, its dimensions also remain as true dimensions. The back of this object, the rear view, is parallel to the front view and so it is also in its true dimensions. This means, then, that the thickness of the object is shown in its true dimensions. This thickness dimension appears in the top and right side views.

The top and bottom surfaces are now oblique to the horizontal plane of projection and so the lengthwise dimensions in the top view are all foreshortened. In other words, any surface that is no longer parallel to its plane of projection will have foreshortened lines. For example, the side surfaces are oblique to their plane of projection and so the vertical dimensions in the right side view are all foreshortened.

Before leaving Fig. 9-2, consider the results of the revolution. The tilt did not affect the front view's dimensions, for no matter what the angle of tilt the front view always remains parallel to the projection plane in front of it. We could have tilted the object by any amount — 45°, 75° or 90°. But as long as we revolved the object on an axis perpendicular to the plane and horizontal to the front view, the true dimensions of the front view remain unchanged. That is not the situation with respect to the other views.

We can take the same block that we have in Fig. 9-2 and revolve it on an axis perpendicular to the horizontal plane of projection. You can see the results in Fig. 9-3A. Note that the top view is unchanged, except for the 30° angle of tilt, and all dimensions in the top view

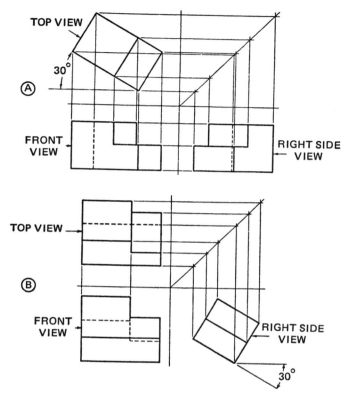

Fig. 9-3. (A). Revolution of top view. (B). Revolution of right side view.

are true. Finally, as illustrated in Fig. 9-3B, we have revolved the object on an axis perpendicular to the profile plane. The right side view has true dimensions.

Advantage Of A Revolved View

The advantage of the usual orthographic projection is that you have true dimensions in all three planes — top, front and side. The disadvantage is that in real life we seldom examine an object in this way. We are accustomed to looking at things in three dimensions simultaneously. An orthographic drawing does supply three dimensions, but since it furnishes only one dimension at a time we may have some difficulty in visualizing just what the object looks like.

When we revolve an object, we sacrifice dimension accuracy. Some lines will be foreshortened. However, it is possible that all of the information we need can be supplied by a single view.

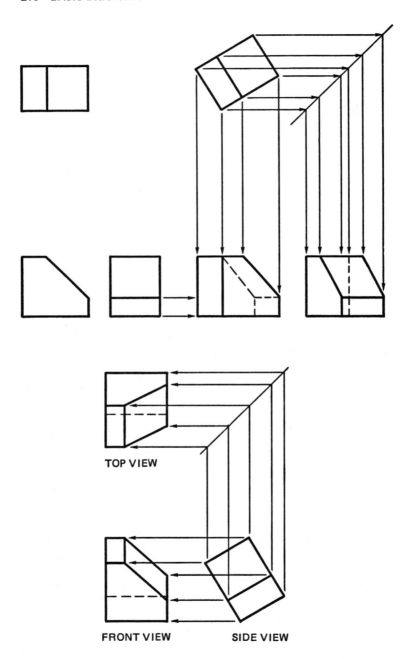

TOP VIEW

FRONT VIEW   **SIDE VIEW**

Fig. 9-4. Regular orthographic projection (top left). Top view revolved (top right) and side view revolved (lower drawing).

In Figs. 9-2 and 9-3 we do not have a concept of three dimensions, but we can get this effect by including hidden lines. In Fig. 9-4 we have an object with a slant edge. This is a rather simple object and so the orthographic projection shown at the top left may be satisfactory. However, if we revolve the top view and include dashed lines to represent hidden views, we can get a more accurate impression of what the object looks like.

By revolving another view — the side view as shown in the lower drawing — and by using dashed lines to indicate hidden views, we can get still another look at the object. Thus, by revolving each view in turn, the effect will be as though we had the object in our hands, revolving the object to get a better look at it. It is often difficult to decide which of the orthographic views to revolve. By making quick sketches instead of working with a drawing board, triangle and T-square, you may be able to select the correct view for revolution the first time.

Numbering Technique

We can use a numbering technique in connection with orthographic projection to help in the revolution of a view. Figure 9-5A shows the object in the upper right corner. Each corner of the object has a number. Carry these numbers over to the orthographic projection. If you will examine the top view of the orthographic projection, you will see numbers 5-8, 6-7, 1-4 and 2-3. In these number pairs the first digit is the corner you can see; the second digit is the hidden view. Thus, in the top view, 1 represents a visible corner while digit 4 is the hidden or invisible corner. Digit 5 is a visible corner while digit 8 is the hidden corner.

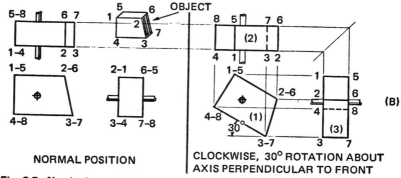

(A) NORMAL POSITION

(B) CLOCKWISE, 30° ROTATION ABOUT AXIS PERPENDICULAR TO FRONT

Fig. 9-5. Numbering technique as an aid in drawing a revolved object. (A). Each corner of the object has a number (B). Clockwise rotation through an angle of 30° of the front view and the top and side views obtained from this rotated view.

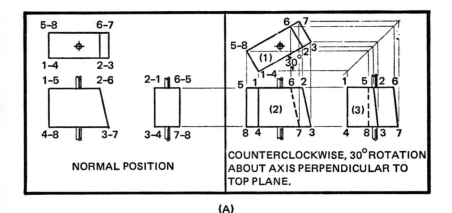

(A)

CLOCKWISE, 30° ROTATION ABOUT
AXIS PERPENDICULAR TO SIDE PLANE.

(B)

Fig. 9-6. (A). Revolution of the top view. (B). Revolution of the side view

Note that the pictorial in the right corner does not show all the numbers. Digit 8 is missing because 8 is a hidden corner in this drawing. With the help of these numbers you can help identify visible and invisible corners in all three of the orthographic views.

Figure 9-5B shows the clockwise rotation through an angle of 30° of the front view and the top and side views obtained from this rotated view. With the help of the numbers you can identify each of the corners of the object.

You can use the same numbering technique no matter which view you decide to turn. In Figs. 9-6A and 9-6B, identification numbers are placed near the rotated top view and also alongside the corners in the rotated side view.

In these examples a line will always have the same numbers at its respective ends in all views. A surface will always have the same numbers at its corresponding corners in all views. To draw its shape, join the points in the same sequence in all views. Remember also that lines which are parallel in one view will also be parallel in the other views.

Successive Revolutions

In Figs. 9-5 and 9-6 our technique was to select a particular view, such as the top, front or side views, and then to rotate the selected view by 30°. We would then project into the other two views from the rotated view. While this method, plus the use of hidden lines, does give us an excellent concept of the object being studied, we can get additional views by using successive revolutions.

As an example of this method, shown in Fig. 9-7, we are going to start with the usual orthographic projection. The object being drawn is illustrated in the upper right corner of Fig. 9-7A.

The object has numbers at each corner and the orthographic projection is numbered accordingly. Note that corner 2 is not seen in the pictorial but does appear in the projection.

Our first step is to rotate the front view by 30°. After making this revolution we project from the front to the top and side views. Note in Fig. 9-7B that all the corner numbers have been carried along. You would, of course, erase all numbers after completing the drawing. In Fig. 9-7B we started with the front view, then projected the top view, and finally the side view. The front view was rotated 30° about an axis perpendicular to the front plane. To help you visualize this, look at points 4, 3 in the front view in the orthographic projection. Put a pencil on these points and hold the pencil vertically. This pencil will represent the axis around which the front view will rotate.

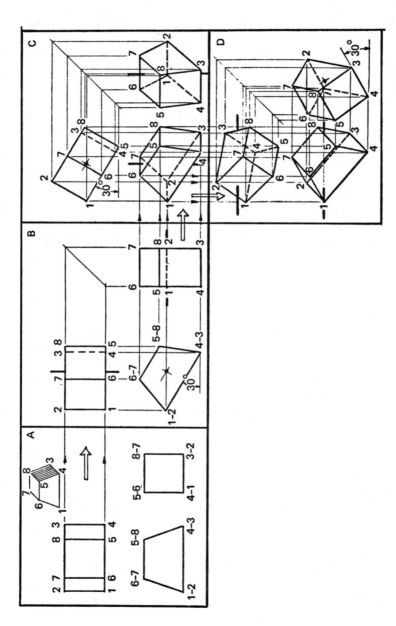

Fig. 9-7. Successive revolutions of an object. (A). Typical orthographic projection. (B). The revolution of the top view. (D). The revolution of the side view. (C). The revolution of the front view.

The front view will turn in a clockwise direction, with the pencil as a sort of axle.

Examine Fig. 9-7C. We have taken the top view of Fig. 9-7B and have rotated it by 30°. This additional rotation was around an axis perpendicular to the top plane. In Fig. 9-7B we did the bottom view first, then the front view and finally the side view. As you can see, Fig. 9-7C is a bit more complicated than Fig. 9-7B. In turn, Fig. 9-7B is not as simple as Fig. 9-7A. Successive revolutions mean more lines of projection.

We can have still another revolution by rotating the side view by 30° as shown in Fig. 9-7D. The side view is drawn first, then the front view and finally the top view. Note that the front view in Fig. 9-7B is the same size and shape as the front view in Fig. 9-7A. The top view in Fig. 9-7C is the same size and shape as the top view in Fig. 9-7B.

As you can see, by the time we reach Fig. 9-7D, we have quite a few lines of projection. These are important only in enabling you to draw the different views, but once you have drawn the views they contribute nothing to the drawing. To minimize confusion, draw projection lines lightly, using very thin lines. The lines connecting the corners of the object, edge lines, should be stronger to avoid confusion with projection lines.

True View By Revolution

If an object has a slant surface and that surface is not parallel to any of the three planes of projection, we can still get a true view of the slant surface through successive revolutions. Sometimes this can be done by a single revolution. Generally, in drawings, views by revolution are used when it is necessary to show the true shape of a surface.

If you have a slant surface and want a true view, the first step is to rotate the object so that an edge of the slant surface is parallel to one of the planes of projection. You may need to rotate this edge either horizontally or vertically. This single rotation may or may not make the entire slant surface parallel to a plane of projection. If not, then rotate once again. If the first rotation was in the vertical plane, then the second will be in the horizontal plane. Conversely, with the first rotation in the horizontal plane, the second will be in the vertical plane.

Figure 9-8 illustrates the technique. Figure 9-8A is the usual orthographic projection. In Fig. 9-8B we rotate the top view counterclockwise about an axis perpendicular to the top plane until edge 3–4 is

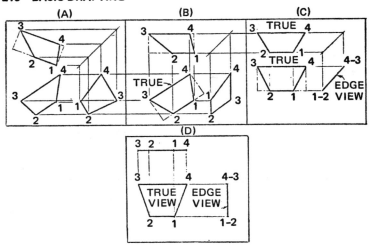

Fig. 9-8. Successive revolutions can be used to get a true view of a slant surface. (A). Orthographic projection. (B). The top view is rotated counterclockwise about an axis perpendicular to the front plane until edge 3-4 is parallel to the top plane. (C). A clockwise rotation is then made. (D). The edge view is rotated until it is parallel to the front plane.

parallel to the front plane. In Fig. 9-8C we rotate clockwise about an axis perpendicular to the front plane until 3-4 is parallel to the top plane. This gives us an edge view of the surface. Finally, in Fig. 9-8D we rotate the edge view until it is parallel to the front plane. The front view is a true view.

Practice Exercise

Figure 9-9 is the drawing of a rectangularly shaped object. Two orthographic views are shown. Draw the third view. Rotate one of the views and then draw top, front and side projections.

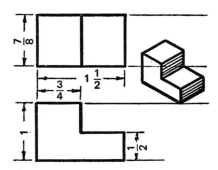

Fig. 9-9. Practice problem. Draw the remaining orthographic view. Revolve one of the views and then draw the projections.

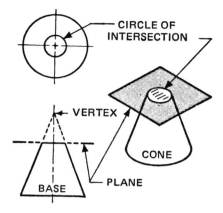

Fig. 9-10. Cutting plane is parallel to the base. Action of the cutting plane produces a circle of intersection. Parallel cutting plane may be any distance above the base.

INTERSECTIONS IN DRAFTING

Objects that are drawn by draftsmen can not only be separate or individual objects but may somehow be joined to another. Typical examples could be a pair of pipes that come together, or a geometric shape such as a cube that is part of another shape such as a cylinder. However, you do not need to have two solids joining to have an intersection. You can have the equivalent effect by having a plane cutting through a solid.

Figure 9-10 shows a cone that is truncated or cut by a cutting plane. In this example the plane is an imaginary device we use for cutting off part of the cone. However, instead of using a cutting plane, the plane itself could represent the base of some other object, such as a cube. In Fig. 9-10 the cutting plane is parallel to the base of the cone. The cutoff portion of the cone is a circle.

Instead of having the cutting plane parallel to the base it could also be vertical to the base, or, if you wish to think of it that way, as parallel to the vertical axis of the cone. The vertical axis is a line drawn from the vertex of the cone, through the center of the cone and vertical to the base. The cutting plane, in removing part of the cone, produces a curve known as a *hyperbola*. The hyperbola is the curved edge of the cutaway section of the cone. See Fig. 9-11.

We can also send the cutting plane through the cone so it is parallel to the slant edge of the cone. The slant edge is called an *element* of the cone. The curved edge that is produced as a result of the cutting action of the plane is known as a *parabola*. This is shown in Fig. 9-12.

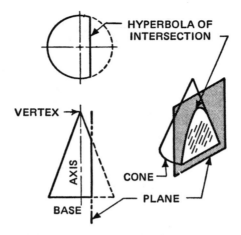

Fig. 9-11. Cutting plane is perpendicular to the base, or may be considered as parallel to the vertical axis of the cone. The cutting action produces a hyperbolic curve. The plane can be any distance from the axis.

Now examine the top views shown in Figs. 9-10, 9-11 and 9-12. In the case of Fig. 9-10, when the cutting plane is sent through the cone, the top view shows the cutoff section as a circle. In Fig. 9-12 the top view reveals the cutoff section as a curve — this time as a parabola. But when the cutting plane is parallel to the axis, as in Fig. 9-11, the cutoff section in the top view appears as a straight line.

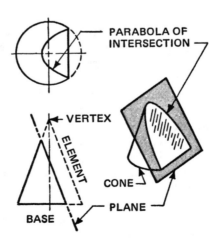

Fig. 9-12. Cutting plane is made parallel to the slant edge or element of the cone. The cutting action produces a parabolic curve. The cutting plane may be any distance from the slant edge of the cone.

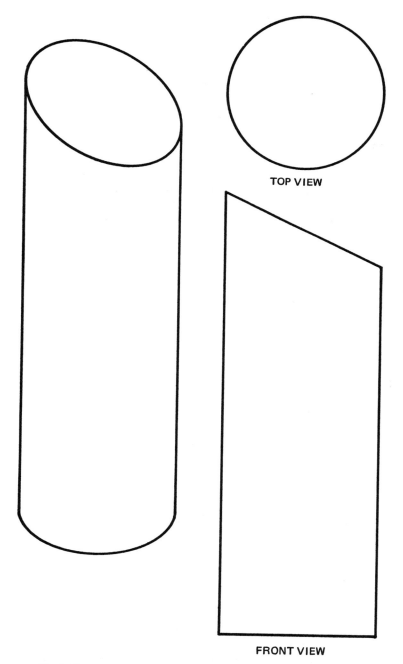

TOP VIEW

FRONT VIEW

Fig. 9-13. Truncated cylinder (left) and top and front views (right).

Figures 9-10, 9-11 and 9-12 all have one thing in common. They all show top and front views, but none of them have a side view. Simply drawing a few projection lines from the top and front views would not produce a satisfactory side view.

Figure 9-13 shows a truncated cylinder and the top and front views. To draw the side view you will need to be able to supply more projection lines. To do this, divide the circle shown in the top view by a number of diagonals (Fig. 9-14). While we show three diagonals in this drawing, a larger number would help produce a more accurate side view. After you have drawn the diagonals, project them downward so they form points of intersection with the front view. Now project lines from the top and front views. The intersections of these lines will supply the points for the formation of the side view (Fig. 9-15). The side view will show an ellipse, and so you will need to use your French curve to draw it.

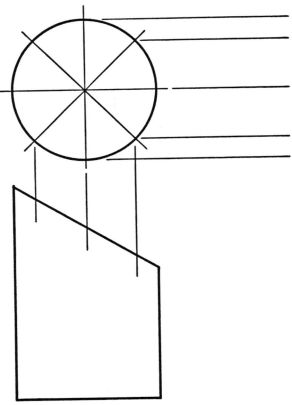

Fig. 9-14. Steps in drawing the side view. Draw a number of diagonals in the top view and then project them vertically and horizontally.

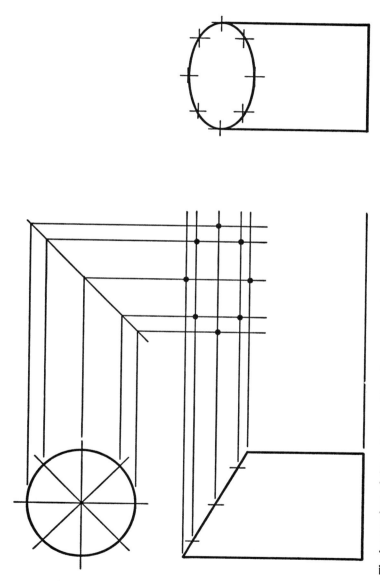

Fig. 9-15. The intersections of the projectors will produce a series of dots. When you connect them you will have the side view. The side view shown at the right has plus symbols (+) to indicate the presence of the dots. When these are connected, the side view will be complete.

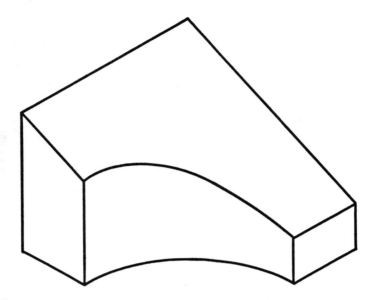

Fig. 9-16. Object which has been cut by a curved cutting plane.

The Curved Cutting Plane

The cutting plane shown in the drawings is a straight line. It could, for example, represent the edge of a box. In some cases it might simply be a geometric shape that has been cut. In the case of the truncated cylinder you can imagine the cutting plane as a knife.

We can also have a curved cutting plane, but this doesn't present any particular problems for we can handle it as though it were straight. Figure 9-16 shows an object which has been sliced by a curved cutting plane.

Figure 9-17 shows the top and front views. The top view reveals that the curved portion is part of a circle. The first step is to locate the center point of that circle. Do this by drawing a number of radii. The point of intersection is shown by the "+" sign. You can then draw vertical projection lines to create the front view. After obtaining the top and front views, project from both of these views to get the side view, shown in Fig. 9-18.

Intersection Of A Prism And Cylinder

Cutting planes are useful in revealing details of an object that may be below its surface. The interior of the object may not necessarily be solid, although externally it may appear to be so. In Fig. 9-19, the drawing at the upper right shows a cylinder intersecting a prism.

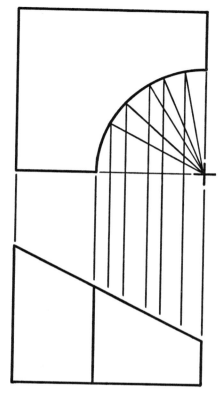

Fig. 9-17. Top and front views of object cut by curved cutting plane.

To draw this combined object we would be interested in knowing more about this intersection. However, this does not prevent us from using a cutting plane to supply more information should we feel this is necessary.

At the upper left in Fig. 9-19 we have the top view of the combined prism-cylinder. The prism appears as a triangle. This is correct since the top face is indeed a triangle. The cylinder, in this view, looks like part of a rectangle.

The front view is illustrated at the lower right. Since we are looking head on at the object, the cylinder will appear to be a circle while the prism will seem to be a square.

Since we now have the top and the front views, we can project the side view. Note that we have transposed the conventional positions of the front and the side views. This was done to show you that it is sometimes more convenient to do so. In some instances this can result in a cleaner projection.

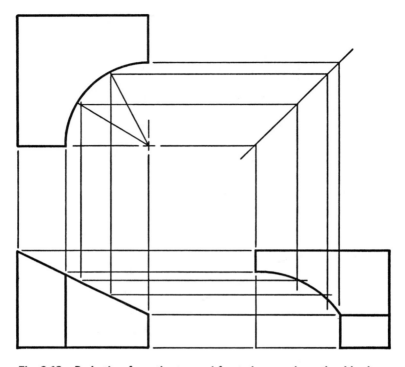

Fig. 9-18. Projection from the top and front views produces the side view.

The intersection points are numbered 1 through 6. These points supply connections for one-half of the curve. Since the curve is symmetrical, the upper half can also be drawn.

One of the dangers in drawing intersections of objects is that it is easy to make a prejudgment about the shape of the curve of intersection. In the case of Fig. 9-19, for example, it would be easy to assume that this would be a circle. But the cylinder appears as a circle only when we look at it from the front view. And, as you can see, the intersection of the cylinder and the prism results in a curve which is not a circle.

What is this intersection of a cylinder and prism? It is a succession of points representing the contacts of two surfaces — that of the cylinder and the prism.

THINGS TO DO

- Assume you have the object shown in Fig. 9-20. Do an orthographic projection. Rotate the front view upward by 30° and then draw the top view, front view and right side view.

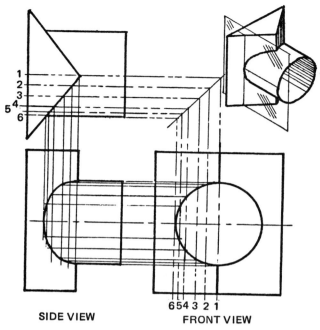

1
2
3
4
5
6

654 3 2 1

SIDE VIEW

FRONT VIEW

Fig. 9-19. Intersection of a cylinder and a prism.

- Figure 9-21 shows an object and its orthographic projection. Redraw the projection to verify the accuracy of the drawing. Rotate the top view by 30°.
- Make an orthographic projection of the object shown in Fig. 9-22. Rotate the top and side views.

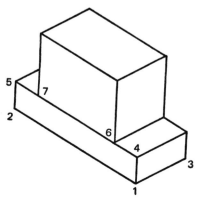

Fig. 9-20. Do an orthographic projection of this object.

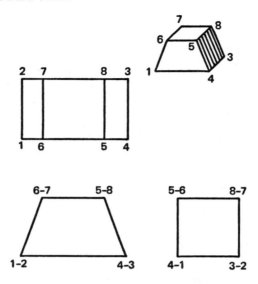

Fig. 9-21. An object and its orthographic projection.

- An orthographic projection is shown in Fig. 9-23. Rotate the bottom view by 30° and then project the other two views. Make a perspective sketch of the object.
- Assume you have a truncated cylinder. Make an orthographic projection of the top, front and side views.

SUMMARY

- If we have an orthographic projection, we can revolve one of the views around an axis which is perpendicular to any one of the planes of projection. Then we draw the object in its new position.

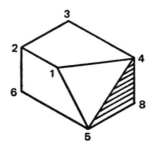

Fig. 9-22. Make an orthographic projection of this object and then rotate the top and side views.

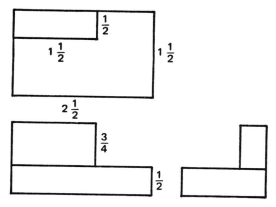

**Fig. 9-23. Another orthographic projection. Rotate the bottom view by 30°
and then project the other two views.**

- When we resolve an object, its dimensions will remain true as
  long as the view remains parallel to the plane of projection.
- If we revolve the front view of an object, its front view dimensions will remain true. The new top and side views of the
  object will not remain true.
- Revolving an object can supply us with a better picture of what
  the object looks like.
- Additional views of an object can be obtained through successive
  revolutions.
- To get a true view of a slant surface, rotate a selected view until
  the slant surface is parallel to one of the planes of projection.
- A cutting plane passed through a cone so it is parallel to the
  slant edge of the cone produces a curve called a parabola.
- When a cutting plane is passed through a cone so that it is
  parallel to the vertical axis of the cone, it produces a curve
  called a hyperbola.
- Cutting planes can be curved or straight.

Glossary

aligned dimensioning — Technique in which all dimensions are placed in line with the dimension lines. Also known as rectangular dimensioning.

ammonia process prints — The transfer of a drawing to a sensitized paper in the presence of ammonia vapor.

architects' scale — A scale used for making measurements on machine, architectural and structural drawings. It contains scales of proportional feet and inches.

auxiliary view — Supplementary view of an object, particularly of a slant surface.

blueprint — Photographic reproduction of a drawing with white lines on a blue background.

bow instruments — Drafting tools for drawing circles with diameters of less than 1 inch and for transferring small measurements.

break lines — Used for indicating an object that is the same throughout its length. A break in a construction line.

brownline — A B–W print with brown lines.

b-w print — The transfer of a drawing to a sensitized paper. Lines of the drawing can be black, brown or maroon.

cabinet drawing — A drawing in which all measurements from front to back are reduced by one-half. Sometimes used in making drawings of wood construction. A drawing using oblique lines of projection.

cavalier drawing — A drawing in which all dimensions shown are true.

center lines — A series of long and short dashes. Used to indicate the center of a symmetrical object.

central projection — Drawing in which all projectors or lines come together at a single point. A drawing in perspective.

chamfer — An edge that has been beveled.

compass — Drafting tool for drawing arcs or circles in pencil or ink.

concentric circles — One or more circles drawn inside another larger circle with all circumference parallel to each other and sharing horizontal and vertical centering lines.

construction lines — Visible outlines.

cross hatching — Series of parallel lines or some specific design for a cross section indicating the kind of material of which the object is made.

cube — Geometric figure with six sides of identical dimensions.

cutaway section — Portion of an object which has been exposed by a cutting plane.

cutting plane lines — Lines used for showing a cross section of an object.

detail paper — Heavy weight opaque paper which is usually buff or neutral green in color.

diameter — Any straight line drawn from one point on the circumference of a circle to any other point on that circumference, passing through the center point of the circle. A diameter is equal to two times the length of a radius.

dimension lines — Lines which end in an arrowhead, or a pair of arrowheads, used in connection with the size of some part of an object.

dividers — Drafting tool for making measurements, transferring measurements or for dividing a line into an equal number of parts.

drafting machine — Combination T-square, drafting triangle and protractor.

drafting media — Any material on which a drawing is made, including paper or cloth.

drawing board — Small, flat drawing surface made of wood. It is portable, and well suited for beginning efforts in drafting.

drop bow pen — Special compass used for drawing circles having a diameter of less than 1/4 inch.

dusting brush — Brush for removing eraser crumbs and dust from a drawing.

engineers' scale — Triangular scale with the scales divided decimally.

erasing machine — Electrically operated device for removing inked or pencil lines from a drawing.

erasing shield — Rectangular plate of metal or plastic with various cutouts to permit erasing small portions of a drawing.

extension lines — Straight thin lines which lead to some part of an object.

fillet — Interior curved angle of an object.

finish marks — Marks used to indicate the kind of finish an object is to have: plated, milled, polished, etc.

foreshortening — A reduction in the length of a line or lines of an object. A line not shown in its true length.

french curve — Drafting tool used as an aid in drawing all types of curved lines.

front view — View of an object looking at it directly from the front.

frontal plane — Plane that receives projectors from the front view of an object.

general notes — Additional information on a drawing relating to all parts of the object.

hidden lines — Lines that represent a hidden part of an object.

horizontal plane — Plane that receives projectors from the top surface of an object.

horizontal projectors — Projectors from the top surface of an object.

india ink — Type of ink used for making drawings.

isometric drawing — Pictorial method that resembles perspective drawing but with its principal dimensions drawn to scale.

isometric projection — A form of projection in which the surfaces of an object are seen in a single view.

leader — Line with an arrowhead and used to call attention to some property of an object.

light table — Drafting table with a glass working surface that is illuminated by lights beneath it.

linen — Cloth made of cotton treated with starch. Can be used with either pencil or ink.

location dimension — Dimension that gives the distance between different parts of the same object.

lofting — Drawings which are the same size as the object. Generally used for large objects.

metric scale — Two bevel scale calibrated in metric units.

microfilm — Strip of standard motion picture film, or smaller, for keeping a photographic record of drawings.

minute — Sixtieth part of a degree.

miter line — Line that divides a 90° angle into two 45° angles.

non-aligned dimensioning — Technique in which dimensions are drawn in all directions.

notes — Supplementary data added to a drawing to supply information concerning certain features of an object.

oblique drawing — A pictorial drawing using oblique lines of projection instead of perpendicular ones.

oblique projection — Drawing in which the projectors form an angle other than 90° with the surfaces of the object.

octagon — Eight-sided geometric figure.

offset section — Series of sections of an object.

orthographic projection — Type of projection in which the lines of projection are perpendicular to the object.

overlay — Sheet of transparent tissue, plastic or cloth placed over a drawing.

parallel projection — Lines projected from some surface of an object which always remain parallel and at right angles to the object. Orthographic projection.

parallel straightedge — Drafting tool controlled by cords that permit up and down movement while keeping the straightedge in a horizontal position.

partial section — Small portion of the interior of an object.

perpendicular point projection — Orthographic projection using a miter line.

phantom lines — Lines for showing an alternate position, or positions, of an object.

photostat — Sometimes called a stat. Photographic reproduction of a drawing. A photostat can be same size, enlarged or reduced. The drawing is made to appear directly on the surface of prepared paper with the image in correct position and not reversed as in a negative.

phototracing — Photographic reproduction of a drawing on tracing cloth.

pictorial drawing — Three-dimensional type of drawing.

plan view — Horizontal projection of an object. Top view.

plane of projection — Imaginary surface at right angles to lines of projection.

primary center line — Main center line in a drawing.

profile — Side view of an object.

profile plane — Plane that receives projectors from the side view of an object.

proportional dividers — Drafting tool for the transfer of measurements from one scale to another. Useful when drawings are to be made to a larger or smaller scale.

protractor – Tool in semicircular or circular form, made of metal or plastic, for measuring angles.

quadrant – One-fourth of a circle. A 90° section of a circle.

radial point projection – Technique used in orthographic projection.

radius – Shortest straight line drawn from the center point of a circle to the circumference. A radius is equal to one-half of a diameter.

rectangle – Four-sided plane figure having two pairs of parallel sides with one pair longer than the other. Each side forms a right angle with connecting sides.

rectangular dimensioning – Aligned dimensioning.

ref – Reference dimension. A dimension that supplies additional information, other than size and location dimensions.

revolution – Revolving an object through a selected number of degrees.

revolved view – View of an object that has been turned a selected number of degrees.

right angle projection – Orthographic projection.

round – An outside curve of an object.

section – A part of an object. Symmetrical object that is cut into two or more parts.

sectioning lines – Lines for showing an exposed surface.

side view – View of an object looking at it directly from the side. Sometimes called the profile or profile view.

size dimension – Dimension that gives the numeric value of a diameter, a width, length or radius of an arc. A size dimension tells how big an object is, or some part of it.

specific notes – Additional information on a drawing relating to part of the object only.

square – Four-sided plane figures having two pairs of parallel sides with all sides of equal dimensions. Each side forms a right angle with connecting sides.

stacked dimensions – Technique in which dimensions are arranged vertically aligned.

staggered dimensioning – Technique in which none of the dimensions are drawn along the same horizontal line.

stat – Abbreviation for photostat.

straightedge – Drafting tool for drawing straight lines. The T-square is used as a straightedge.

template – Device made of plastic or metal containing outlines of symbols, such as architectural, plumbing, electrical and electronic.

throughhole – A hole that is cut completely through an object.

title block – Outlined rectangle in lower right-hand corner of a drawing and containing general notes.

tolerance – Permitted deviation from the size of an object or some part of it.

top view – View of an object obtained by looking directly down on it. Sometimes called the plan view.

tracing cloth – Finely woven fabric that is coated. One side is glossy, the other side is dull. The dull side is generally used as the drawing surface.

tracing paper – Strong, transparent paper used for making copies of drawings or parts of drawings.

tracing vellum – Tracing paper.

triangles — Drafting tools made of plastic or metal. Triangular in shape, they have three angles totaling 180°. One of the angles is always a right angle (90°). The other two angles are 30° and 60° or a pair of 45° angles.

true dimension — Measurement which represents the actual or proportional size of some part of an object.

true view — View showing actual dimensions of an object.

truncation — Slicing of an object or cutting away of a portion of an object.

T-square — Drafting tool used in conjunction with a work surface. It has a long strip called the blade at right angles to a shorter strip called the head.

unidirectional dimensioning — Technique in which all dimensions are placed horizontally.

vellum — Also known as tracing paper or tracing vellum. Paper may be white or slightly tinted.

visible line — Lines that can be seen when viewing an object. Lines that show the outline of an object.

weight of line — Line thickness.

work surface — A drafting board, drafting table or surface of a drafting machine.

X axis — Vertical line used as a reference.

Y axis — Horizontal line used as a reference.

Appendix

Table A-1. Decimal equivalents of fractional parts of an inch.

1/64——.015625		33/64——.515625	
1/32——.03125		17/32——.53125	
3/64——.046875		35/64——.546875	
1/16——.0625		9/16——.5625	
5/64——.078125		37/64——.578125	
3/32——.09375		19/32——.59375	
7/64——.109375		39/64——.609375	
1/8——.1250		5/8——.6250	
9/64——.140625		41/64——.640625	
5/32——.15625		21/32——.65625	
11/64——.171875		43/64——.671875	
3/16——.1875		11/16——.6875	
13/64——.203125		45/64——.703125	
7/32——.21875		23/32——.71875	
15/64——.234375		47/64——.734375	
1/4——.250		3/4——.750	
17/64——.265625		49/64——.765625	
9/32——.28125		25/32——.78125	
19/64——.296875		51/64——.796875	
5/16——.3125		13/16——.8125	
21/64——.328125		53/64——.828125	
11/32——.34375		27/32——.84375	
23/64——.359375		55/64——.859375	
3/8——.3750		7/8——.8750	
25/64——.390625		57/64——.890625	
13/32——.40625		29/32——.90625	
27/64——.421875		59/64——.921875	
7/16——.4375		15/16——.9375	
29/64——.453125		61/64——.953125	
15/32——.46875		31/32——.96875	
31/64——.484375		63/64——.984375	
1/2——.500		1——1.000	

Table A-2. Numbers and their reciprocals.

n	1/n	n	1/n	n	1/n
0.1	10.0000	28	.0357	65	.0154
0.2	5.0000	29	.0345	66	.0152
0.3	3.3333	30	.0333	67	.0149
0.4	2.5000	31	.0323	68	.0147
0.5	2.0000	32	.0313	69	.0145
0.6	1.6666	33	.0303	70	.0143
0.7	1.4286	34	.0294	71	.0141
0.8	1.2500	35	.0286	72	.0139
0.9	1.1111	36	.0278	73	.0137
		37	.0270	74	.0135
1	1.0000	38	.0263	75	.0133
2	.5000	39	.0256	76	.0132
3	.3333	40	.0250	77	.0130
4	.2500	41	.0244	78	.0128
5	.2000	42	.0238	79	.0127
6	.1667	43	.0233	80	.0125
7	.1429	44	.0227	81	.0123
8	.1250	45	.0222	82	.0122
9	.1111	46	.0217	83	.0120
10	.1000	47	.0213	84	.0119
11	.0909	48	.0208	85	.0118
12	.0833	49	.0204	86	.0116
13	.0769	50	.0200	87	.0115
14	.0714	51	.0196	88	.0114
15	.0667	52	.0192	89	.0112
16	.0625	53	.0189	90	.0111
17	.0588	54	.0185	91	.0110
18	.0555	55	.0182	92	.0109
19	.0526	56	.0179	93	.0108
20	.0500	57	.0175	94	.0106
21	.0476	58	.0172	95	.0105
22	.0455	59	.0169	96	.0104
23	.0435	60	.0167	97	.0103
24	.0417	61	.0164	98	.0102
25	.0400	62	.0161	99	.0101
26	.0385	63	.0159	100	.0100
27	.0370	64	.0156		

Table A-3. Square roots of numbers

n	$\sqrt{n}$	n	$\sqrt{n}$	n	$\sqrt{n}$	n	$\sqrt{n}$
1	1.0000	26	5.0990	51	7.1414	76	8.7178
2	1.4142	27	5.1962	52	7.2111	77	8.7750
3	1.7321	28	5.2915	53	7.2801	78	8.8318
4	2.0000	29	5.3852	54	7.3485	79	8.8882
5	2.2361	30	5.4772	55	7.4162	80	8.9443
6	2.4495	31	5.5678	56	7.4833	81	9.0000
7	2.6458	32	5.6569	57	7.5498	82	9.0554
8	2.8284	33	5.7446	58	7.6158	83	9.1104
9	3.0000	34	5.8310	59	7.6811	84	9.1652
10	3.1623	35	5.9161	60	7.7460	85	9.2195
11	3.3166	36	6.0000	61	7.8102	86	9.2736
12	3.4641	37	6.0828	62	7.8740	87	9.3274
13	3.6056	38	6.1644	63	7.9373	88	9.3808
14	3.7417	39	6.2450	64	8.0000	89	9.4340
15	3.8730	40	6.3246	65	8.0623	90	9.4868
16	4.0000	41	6.4031	66	8.1240	91	9.5394
17	4.1231	42	6.4807	67	8.1854	92	9.5917
18	4.2426	43	6.5574	68	8.2462	93	9.6437
19	4.3589	44	6.6332	69	8.3066	94	9.6954
20	4.4721	45	6.7082	70	8.3666	95	9.7468
21	4.5826	46	6.7823	71	8.4261	96	9.7980
22	4.6904	47	6.8557	72	8.4853	97	9.8490
23	4.7958	48	6.9282	73	8.5440	98	9.8995
24	4.8990	49	7.0000	74	8.6023	99	9.9499
25	5.0000	50	7.0711	75	8.6603	100	10.0000

Table A-4. Diameter, circumference and area of circles.

Diameter	Circumference	Area
1/32	0.09817	0.0007
1/16	0.19635	0.0030
3/32	0.29452	0.0069
3/16	0.58904	0.0276
7/32	0.68722	0.0375
9/32	0.88357	0.0621
11/32	1.07992	0.0928
13/32	1.27627	0.1296
9/16	1.76715	0.2485
19/32	1.86532	0.2768
21/32	2.06167	0.3382
11/16	2.15984	0.3712
23/32	2.25802	0.4057
25/32	2.45437	0.4793
27/32	2.65072	0.5591
29/32	2.84707	0.6450
1	3.142	0.7854
2	6.283	3.1416
3	9.425	7.0686
4	12.566	12.5664
5	15.708	19.6350
6	18.850	28.2743
7	21.991	38.4845
8	25.133	50.2655
9	28.274	63.6173
10	31.416	78.5398
11	34.558	95.0332
12	37.699	113.097
13	40.841	132.732
14	43.982	153.938
15	47.124	176.715
16	50.265	201.062
17	53.407	226.980
18	56.549	254.469
19	59.690	283.529
20	62.832	314.159

Table A-4. Diameter, circumference and area of circles (cont.).

Diameter	Circumference	Area
21	65.973	346.361
22	69.115	380.133
23	72.257	415.476
24	75.398	452.389
25	78.540	490.874
26	81.681	530.929
27	84.823	572.555
28	87.965	615.752
29	91.106	660.520
30	94.248	706.858
31	97.389	754.768
32	100.531	804.248
33	103.673	855.299
34	106.814	907.920
35	109.956	962.113
36	113.097	1,017.88
37	116.239	1,075.21
38	119.381	1,134.11
39	122.522	1,194.59
40	125.66	1,256.64
41	128.81	1,320.25
42	131.95	1,385.44
43	135.09	1,452.20
44	138.23	1,520.53
45	141.37	1,590.43
46	144.51	1,661.90
47	147.65	1,734.94
48	150.80	1,809.56
49	153.94	1,885.74
50	157.08	1,963.50
51	160.22	2,042.82
52	163.36	2,123.72
53	166.50	2,206.18
54	169.65	2,290.22
55	172.79	2,375.83

Table A-4. Diameter, circumference and area of circles (cont.).

Diameter	Circumference	Area
56	175.93	2,463.01
57	179.07	2,551.76
58	182.21	2,642.08
59	185.35	2,733.97
60	188.50	2,827.43
61	191.64	2,922.47
62	194.78	3,019.07
63	197.92	3,117.25
64	201.06	3,216.99
65	204.20	3,318.31
66	207.35	3,421.19
67	210.49	3,525.65
68	213.63	3,631.68
69	216.77	3,739.28
70	219.91	3,848.45
71	223.05	3,959.19
72	226.19	4,071.50
73	229.34	4,185.39
74	232.48	4,300.84
75	235.62	4,417.86
76	238.76	4,536.46
77	241.90	4,656.63
78	245.04	4,778.36
79	248.19	4,901.67
80	251.33	5,026.55
81	254.47	5,153.00
82	257.61	5,281.02
83	260.75	5,410.61
84	263.89	5,541.77
85	267.04	5,674.50
86	270.18	5,808.80
87	273.32	5,944.68
88	276.46	6,082.12
89	279.60	6,221.14
90	282.74	6,361.73

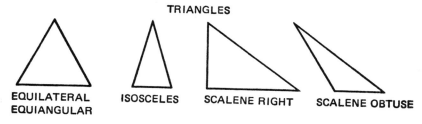

Fig. A-1. Triangles used in drafting.

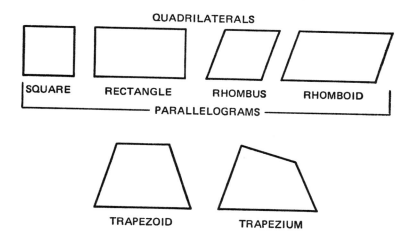

Fig. A-2. Quadrilaterals used in drafting.

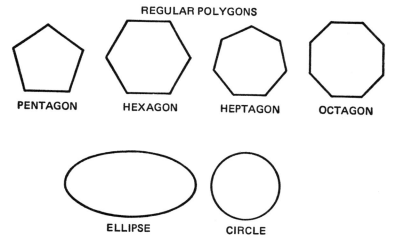

Fig. A-3. Regular polygons used in drafting.

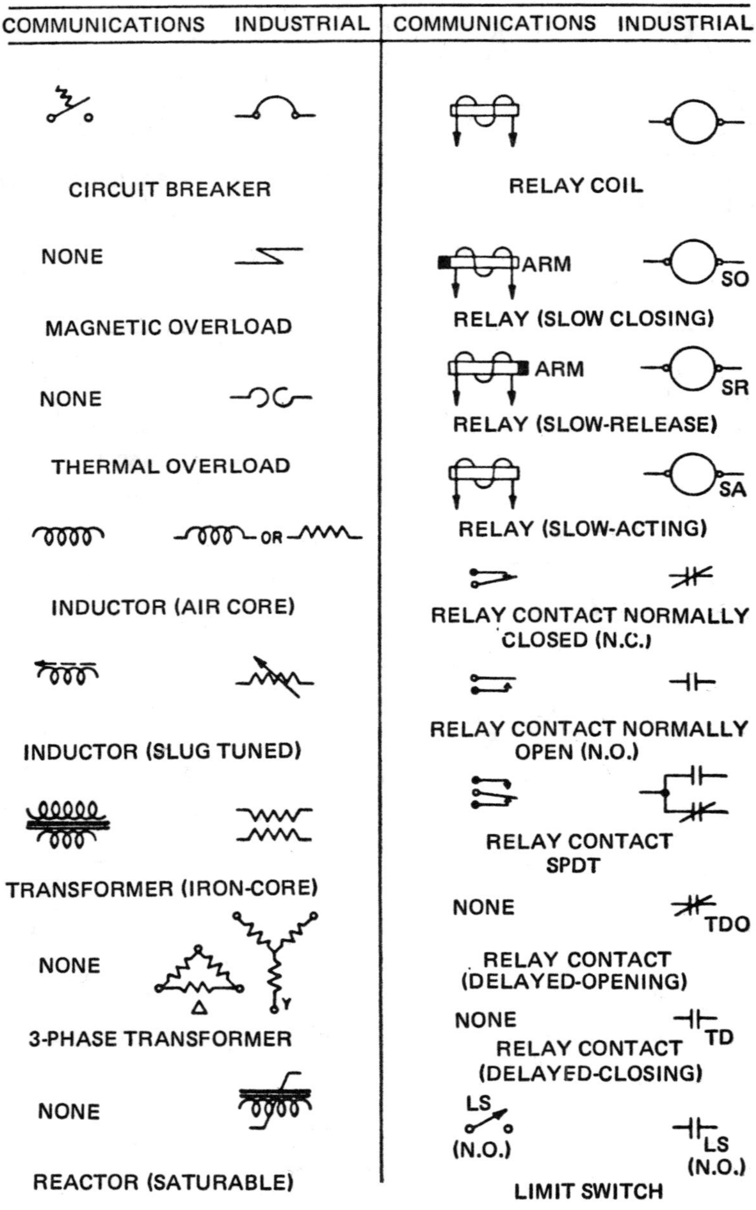
COMMUNICATIONS
INDUSTRIAL
COMMUNICATIONS
INDUSTRIAL
CIRCUIT BREAKER
NONE
MAGNETIC OVERLOAD
NONE
THERMAL OVERLOAD
OR
INDUCTOR (AIR CORE)
INDUCTOR (SLUG TUNED)
TRANSFORMER (IRON-CORE)
NONE
Δ
Y
3-PHASE TRANSFORMER
NONE
REACTOR (SATURABLE)
RELAY COIL
ARM
SO
RELAY (SLOW CLOSING)
ARM
SR
RELAY (SLOW-RELEASE)
SA
RELAY (SLOW-ACTING)
RELAY CONTACT NORMALLY CLOSED (N.C.)
RELAY CONTACT NORMALLY OPEN (N.O.)
RELAY CONTACT SPDT
NONE
TDO
RELAY CONTACT (DELAYED-OPENING)
NONE
TD
RELAY CONTACT (DELAYED-CLOSING)
LS
(N.O.)
LS
(N.O.)
LIMIT SWITCH

Fig. A-4. Schematic symbols used in electronics.

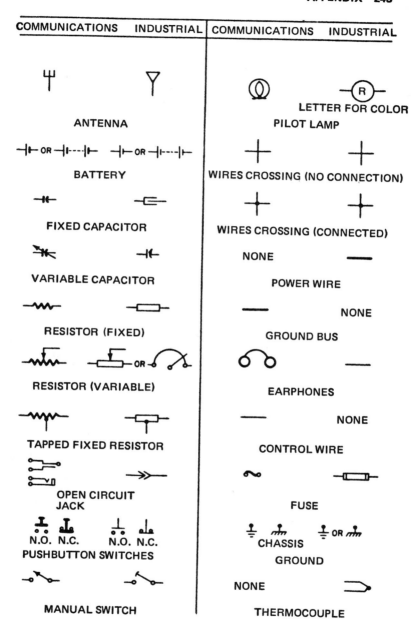

Fig. A-4. Schematic symbols used in electronics. (cont'd.)

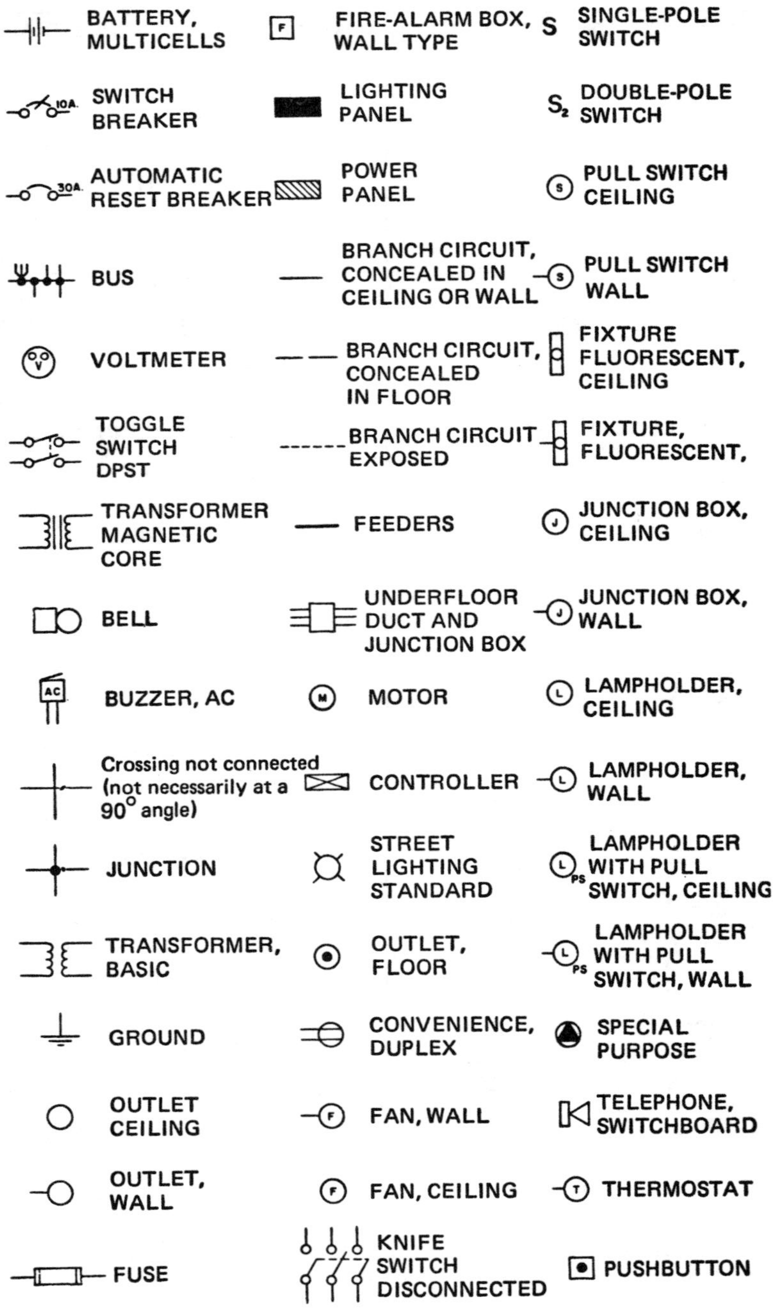

⊣⊢⊢ BATTERY, MULTICELLS	F FIRE-ALARM BOX, WALL TYPE	S SINGLE-POLE SWITCH
⊶⊶ 10A. SWITCH BREAKER	▬▬ LIGHTING PANEL	S₂ DOUBLE-POLE SWITCH
⊶⊶ 30A. AUTOMATIC RESET BREAKER	▨ POWER PANEL	Ⓢ PULL SWITCH CEILING
⊻⊦⊦ BUS	— BRANCH CIRCUIT, CONCEALED IN CEILING OR WALL	⊸Ⓢ PULL SWITCH WALL
☉ VOLTMETER	– – BRANCH CIRCUIT, CONCEALED IN FLOOR	▤ FIXTURE FLUORESCENT, CEILING
⊶⊶ TOGGLE SWITCH DPST	– – – – BRANCH CIRCUIT EXPOSED	⊟ FIXTURE, FLUORESCENT,
⊰⊱ TRANSFORMER MAGNETIC CORE	— FEEDERS	Ⓙ JUNCTION BOX, CEILING
▭◯ BELL	⊟ UNDERFLOOR DUCT AND JUNCTION BOX	⊸Ⓙ JUNCTION BOX, WALL
⊡ BUZZER, AC	Ⓜ MOTOR	Ⓛ LAMPHOLDER, CEILING
⊣ Crossing not connected (not necessarily at a 90° angle)	⊠ CONTROLLER	⊸Ⓛ LAMPHOLDER, WALL
⊣ JUNCTION	☿ STREET LIGHTING STANDARD	Ⓛₚₛ LAMPHOLDER WITH PULL SWITCH, CEILING
⊰⊱ TRANSFORMER, BASIC	⊙ OUTLET, FLOOR	⊸Ⓛₚₛ LAMPHOLDER WITH PULL SWITCH, WALL
⊥ GROUND	⊖ CONVENIENCE, DUPLEX	◉ SPECIAL PURPOSE
◯ OUTLET CEILING	⊸Ⓕ FAN, WALL	◁ TELEPHONE, SWITCHBOARD
⊸◯ OUTLET, WALL	Ⓕ FAN, CEILING	⊸Ⓣ THERMOSTAT
⊟⊟ FUSE	⚡ KNIFE SWITCH DISCONNECTED	▣ PUSHBUTTON

Fig. A-5. Electrical symbols.

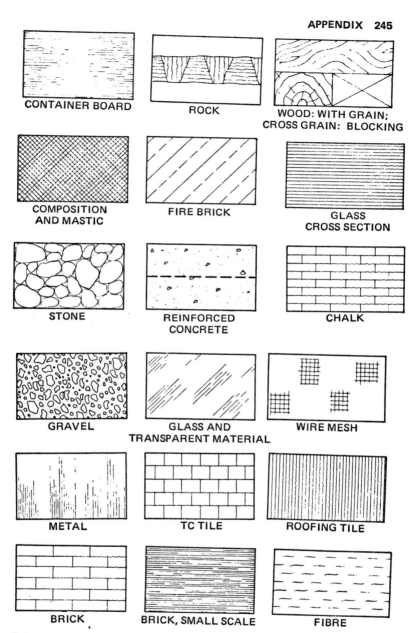

CONTAINER BOARD

ROCK

WOOD: WITH GRAIN;
CROSS GRAIN: BLOCKING

COMPOSITION
AND MASTIC

FIRE BRICK

GLASS
CROSS SECTION

STONE

REINFORCED
CONCRETE

CHALK

GRAVEL

GLASS AND
TRANSPARENT MATERIAL

WIRE MESH

METAL

TC TILE

ROOFING TILE

BRICK

BRICK, SMALL SCALE

FIBRE

✽ INSTEAD OF INDICATING AGGREGATE, SMUDGE ON REVERSE
SIDE OF LINEN

Fig. A-6. Drafting symbols for various materials.

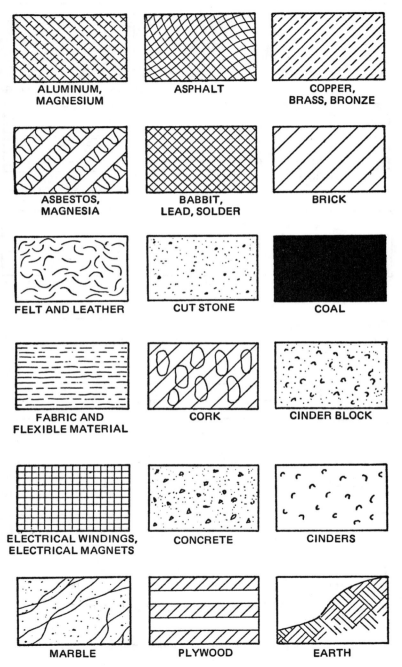

ALUMINUM,
MAGNESIUM

ASPHALT

COPPER,
BRASS, BRONZE

ASBESTOS,
MAGNESIA

BABBIT,
LEAD, SOLDER

BRICK

FELT AND LEATHER

CUT STONE

COAL

FABRIC AND
FLEXIBLE MATERIAL

CORK

CINDER BLOCK

ELECTRICAL WINDINGS,
ELECTRICAL MAGNETS

CONCRETE

CINDERS

MARBLE

PLYWOOD

EARTH

Fig. A-6. Continued

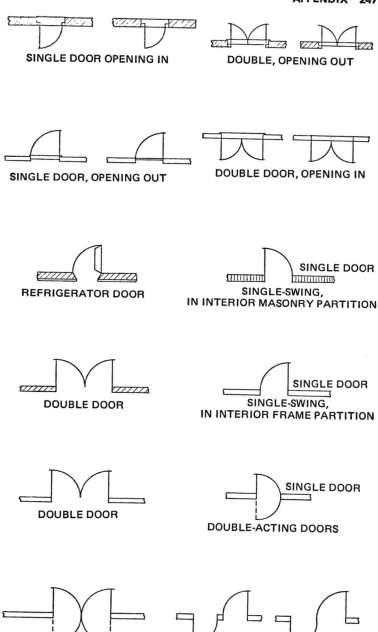

SINGLE DOOR OPENING IN

DOUBLE, OPENING OUT

SINGLE DOOR, OPENING OUT

DOUBLE DOOR, OPENING IN

REFRIGERATOR DOOR

SINGLE DOOR
SINGLE-SWING,
IN INTERIOR MASONRY PARTITION

DOUBLE DOOR

SINGLE DOOR
SINGLE-SWING,
IN INTERIOR FRAME PARTITION

DOUBLE DOOR

SINGLE DOOR
DOUBLE-ACTING DOORS

DOUBLE DOOR

IN AND OUT DOORS

Fig. A-7. Door symbols.

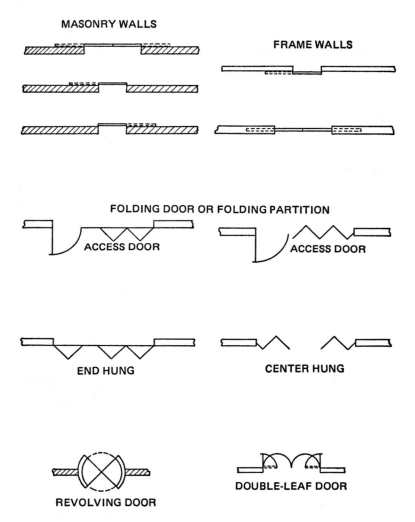

Fig. A-7. Door symbols. (cont'd.)

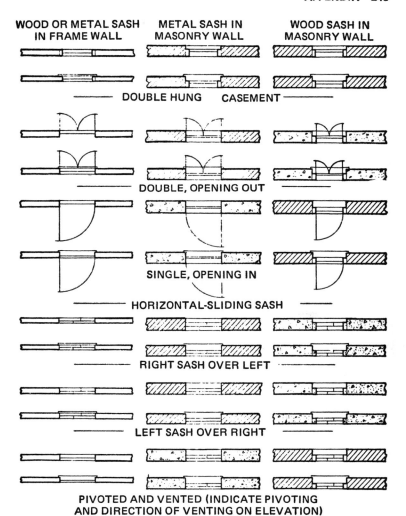

Fig. A-8. Window symbols.

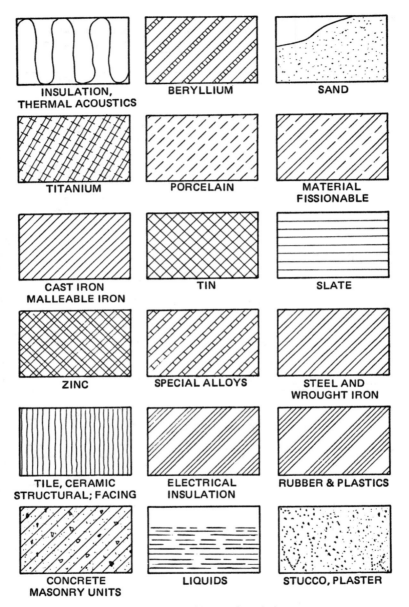

Fig. A-9. Architectural symbols.

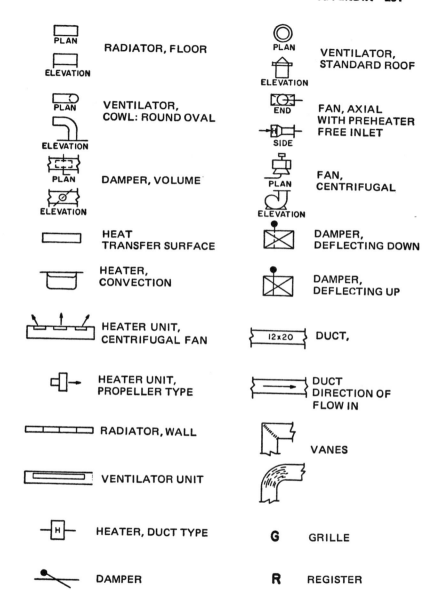

Fig. A-10. Heating symbols.

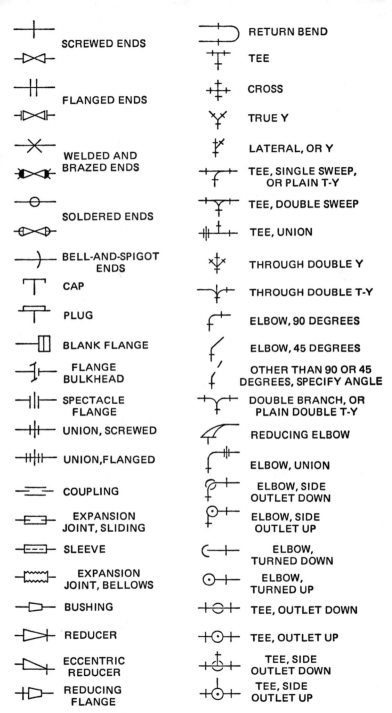

SCREWED ENDS	RETURN BEND
	TEE
FLANGED ENDS	CROSS
	TRUE Y
WELDED AND BRAZED ENDS	LATERAL, OR Y
	TEE, SINGLE SWEEP, OR PLAIN T-Y
SOLDERED ENDS	TEE, DOUBLE SWEEP
	TEE, UNION
BELL-AND-SPIGOT ENDS	THROUGH DOUBLE Y
CAP	THROUGH DOUBLE T-Y
PLUG	ELBOW, 90 DEGREES
BLANK FLANGE	ELBOW, 45 DEGREES
FLANGE BULKHEAD	OTHER THAN 90 OR 45 DEGREES, SPECIFY ANGLE
SPECTACLE FLANGE	DOUBLE BRANCH, OR PLAIN DOUBLE T-Y
UNION, SCREWED	REDUCING ELBOW
UNION, FLANGED	ELBOW, UNION
COUPLING	ELBOW, SIDE OUTLET DOWN
EXPANSION JOINT, SLIDING	ELBOW, SIDE OUTLET UP
SLEEVE	ELBOW, TURNED DOWN
EXPANSION JOINT, BELLOWS	ELBOW, TURNED UP
BUSHING	TEE, OUTLET DOWN
REDUCER	TEE, OUTLET UP
ECCENTRIC REDUCER	TEE, SIDE OUTLET DOWN
REDUCING FLANGE	TEE, SIDE OUTLET UP

Fig. A-11. Plumbing symbols.

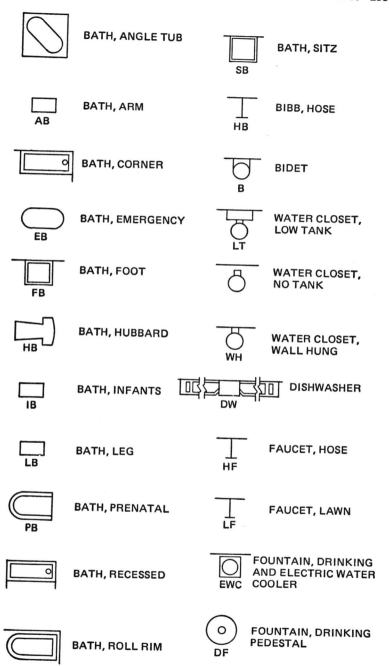

BATH, ANGLE TUB	BATH, SITZ — SB
BATH, ARM — AB	BIBB, HOSE — HB
BATH, CORNER	BIDET — B
BATH, EMERGENCY — EB	WATER CLOSET, LOW TANK — LT
BATH, FOOT — FB	WATER CLOSET, NO TANK
BATH, HUBBARD — HB	WATER CLOSET, WALL HUNG — WH
BATH, INFANTS — IB	DISHWASHER — DW
BATH, LEG — LB	FAUCET, HOSE — HF
BATH, PRENATAL — PB	FAUCET, LAWN — LF
BATH, RECESSED	FOUNTAIN, DRINKING AND ELECTRIC WATER COOLER — EWC
BATH, ROLL RIM	FOUNTAIN, DRINKING PEDESTAL — DF

Fig. A-11. Plumbing symbols. (cont'd.).

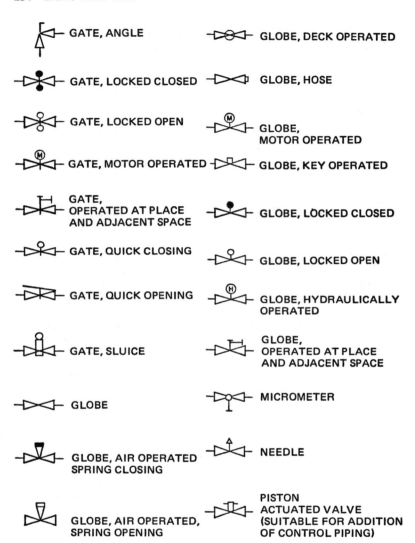

GATE, ANGLE

GLOBE, DECK OPERATED

GATE, LOCKED CLOSED

GLOBE, HOSE

GATE, LOCKED OPEN

GLOBE, MOTOR OPERATED

GATE, MOTOR OPERATED

GLOBE, KEY OPERATED

GATE, OPERATED AT PLACE AND ADJACENT SPACE

GLOBE, LOCKED CLOSED

GATE, QUICK CLOSING

GLOBE, LOCKED OPEN

GATE, QUICK OPENING

GLOBE, HYDRAULICALLY OPERATED

GATE, SLUICE

GLOBE, OPERATED AT PLACE AND ADJACENT SPACE

GLOBE

MICROMETER

GLOBE, AIR OPERATED SPRING CLOSING

NEEDLE

GLOBE, AIR OPERATED, SPRING OPENING

PISTON ACTUATED VALVE (SUITABLE FOR ADDITION OF CONTROL PIPING)

Fig. A-11. Plumbing symbols. (cont'd.).

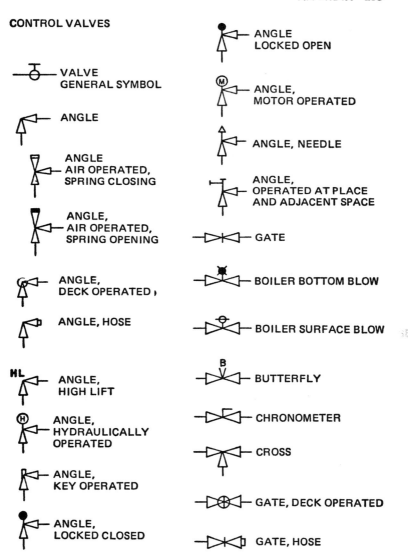

Fig. A-11. Plumbing symbols. (cont'd.).

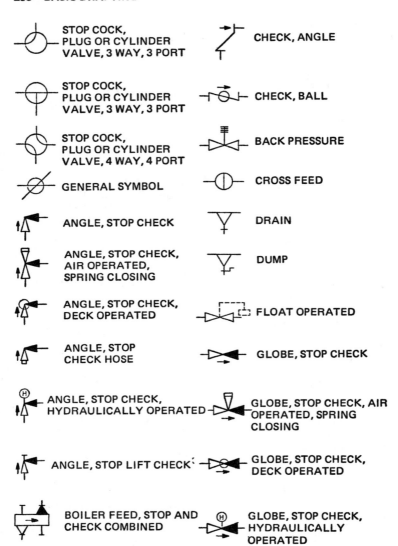

STOP COCK, PLUG OR CYLINDER VALVE, 3 WAY, 3 PORT

STOP COCK, PLUG OR CYLINDER VALVE, 3 WAY, 3 PORT

STOP COCK, PLUG OR CYLINDER VALVE, 4 WAY, 4 PORT

GENERAL SYMBOL

ANGLE, STOP CHECK

ANGLE, STOP CHECK, AIR OPERATED, SPRING CLOSING

ANGLE, STOP CHECK, DECK OPERATED

ANGLE, STOP CHECK HOSE

ANGLE, STOP CHECK, HYDRAULICALLY OPERATED

ANGLE, STOP LIFT CHECK

BOILER FEED, STOP AND CHECK COMBINED

CHECK, ANGLE

CHECK, BALL

BACK PRESSURE

CROSS FEED

DRAIN

DUMP

FLOAT OPERATED

GLOBE, STOP CHECK

GLOBE, STOP CHECK, AIR OPERATED, SPRING CLOSING

GLOBE, STOP CHECK, DECK OPERATED

GLOBE, STOP CHECK, HYDRAULICALLY OPERATED

Fig. A-11. Plumbing symbols. (cont'd.).

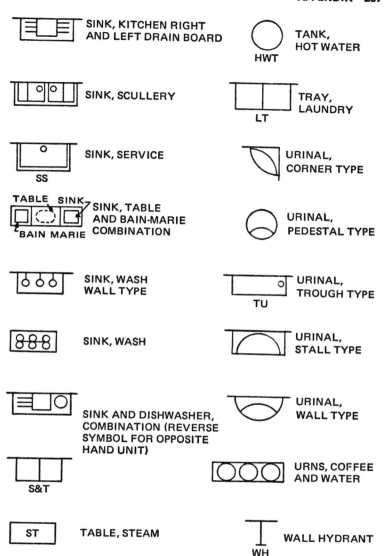

SINK, KITCHEN RIGHT AND LEFT DRAIN BOARD

TANK, HOT WATER
HWT

SINK, SCULLERY

TRAY, LAUNDRY
LT

SINK, SERVICE
SS

URINAL, CORNER TYPE

SINK, TABLE AND BAIN-MARIE COMBINATION

URINAL, PEDESTAL TYPE

SINK, WASH WALL TYPE

URINAL, TROUGH TYPE
TU

SINK, WASH

URINAL, STALL TYPE

SINK AND DISHWASHER, COMBINATION (REVERSE SYMBOL FOR OPPOSITE HAND UNIT)
S&T

URINAL, WALL TYPE

URNS, COFFEE AND WATER

TABLE, STEAM
ST

WALL HYDRANT
WH

Fig. A-11. Plumbing symbols. (cont'd.).

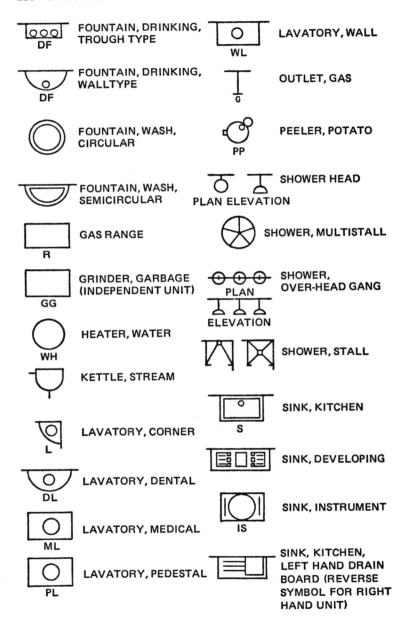

FOUNTAIN, DRINKING, TROUGH TYPE	LAVATORY, WALL
FOUNTAIN, DRINKING, WALLTYPE	OUTLET, GAS
FOUNTAIN, WASH, CIRCULAR	PEELER, POTATO
FOUNTAIN, WASH, SEMICIRCULAR	SHOWER HEAD
GAS RANGE	SHOWER, MULTISTALL
GRINDER, GARBAGE (INDEPENDENT UNIT)	SHOWER, OVER-HEAD GANG
HEATER, WATER	SHOWER, STALL
KETTLE, STREAM	SINK, KITCHEN
LAVATORY, CORNER	SINK, DEVELOPING
LAVATORY, DENTAL	SINK, INSTRUMENT
LAVATORY, MEDICAL	SINK, KITCHEN, LEFT HAND DRAIN BOARD (REVERSE SYMBOL FOR RIGHT HAND UNIT)
LAVATORY, PEDESTAL	

Fig. A-11. Plumbing symbols. (cont'd.).

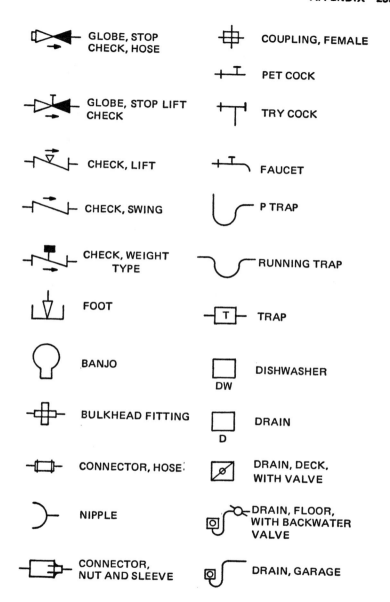

GLOBE, STOP
CHECK, HOSE

COUPLING, FEMALE

PET COCK

GLOBE, STOP LIFT
CHECK

TRY COCK

CHECK, LIFT

FAUCET

CHECK, SWING

P TRAP

CHECK, WEIGHT
TYPE

RUNNING TRAP

FOOT

TRAP

BANJO

DISHWASHER
DW

BULKHEAD FITTING

DRAIN
D

CONNECTOR, HOSE

DRAIN, DECK,
WITH VALVE

NIPPLE

DRAIN, FLOOR,
WITH BACKWATER
VALVE

CONNECTOR,
NUT AND SLEEVE

DRAIN, GARAGE

Fig. A-11. Plumbing symbols. (cont'd.).

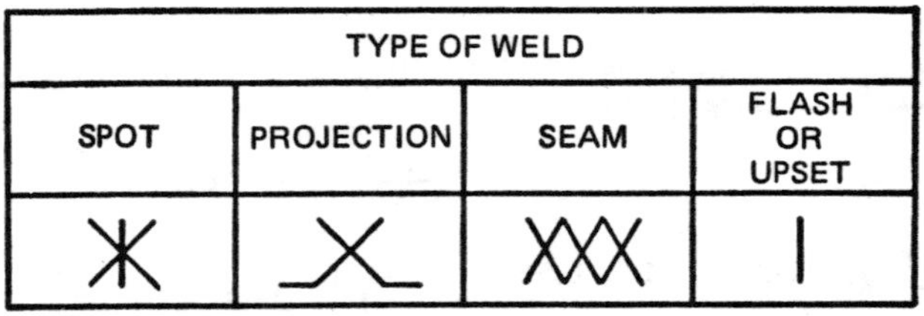

TYPE OF WELD			
SPOT	PROJECTION	SEAM	FLASH OR UPSET
✳	⟍	XXX	│

Fig. A-12. Basic resistance weld symbols.

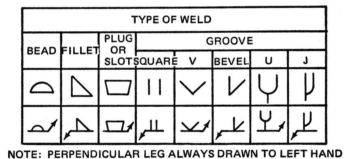

			TYPE OF WELD				
		PLUG OR SLOT	GROOVE				
BEAD	FILLET		SQUARE	V	BEVEL	U	J
⌓	◺	⏢	‖	⋁	⋁	Y	⋃
⌓	◺	⏢	⊥⊥	⋁	⋁	Y	⊍

NOTE: PERPENDICULAR LEG ALWAYS DRAWN TO LEFT HAND

Fig. A-13. Basic arc and gas weld symbols.

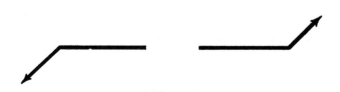

Fig. A-14. Basic welding symbols.

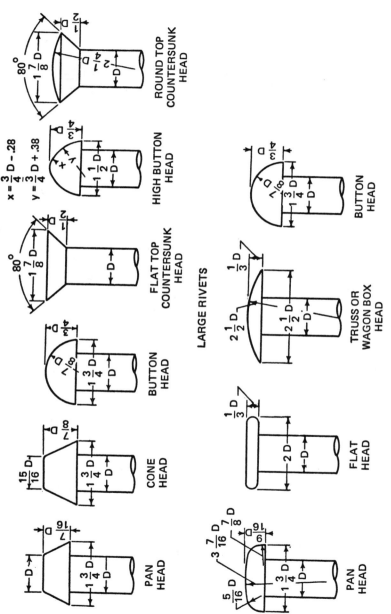

Fig. A-15. Shown are various types of rivets.

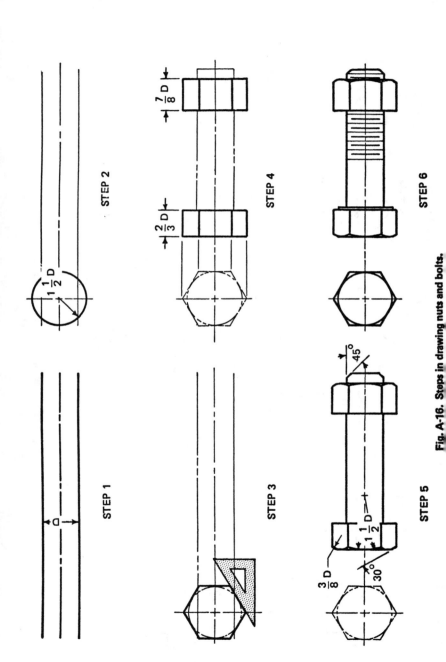

Fig. A-16. Steps in drawing nuts and bolts.

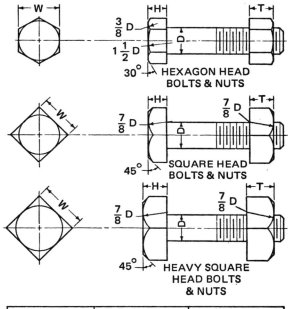

Semi-finished Regular	Unfinished Regular	Unfinished Heavy
$W = 1\frac{1}{2}D$	$W = 1\frac{1}{2}D$	$W = 1\frac{1}{2}D + \frac{1}{8}$
$H = \frac{3}{4}D$	$H = \frac{2}{3}D$	$H = \frac{W}{2}$
$T = \frac{7}{8}D$	$T = \frac{7}{8}D$	$T = D$

Fig. A-17. Bolt and nut formulas.

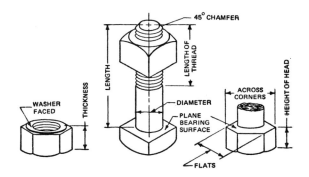

Fig. A-18. Bolt specifications.

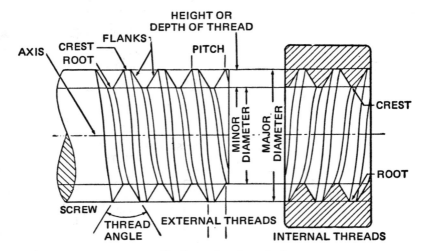

Fig. A-19.   Screw-thread terminology.

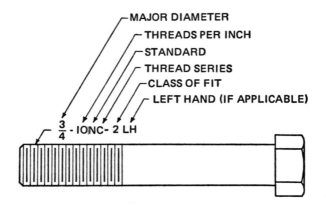

Fig. A-20.   Thread specifications.

Index

Index